Code of Practice for Programme Management in the Built Environment

Code of Practice for Programme Management in the Built Environment

THE CHARTERED INSTITUTE OF BUILDING

WILEY Blackwell

This edition first published 2016
© 2016 by John Wiley & Sons, Ltd

Registered Office
John Wiley & Sons, Ltd, The Atrium, Southern Gate, Chichester, West Sussex, PO19 8SQ, United Kingdom

Editorial Offices
9600 Garsington Road, Oxford, OX4 2DQ, United Kingdom
The Atrium, Southern Gate, Chichester, West Sussex, PO19 8SQ, United Kingdom

For details of our global editorial offices, for customer services and for information about how to apply for permission to reuse the copyright material in this book please see our website at www.wiley.com/wiley-blackwell.

The right of the author to be identified as the author of this work has been asserted in accordance with the UK Copyright, Designs and Patents Act 1988.

All rights reserved. No part of this publication may be reproduced, stored in a retrieval system, or transmitted, in any form or by any means, electronic, mechanical, photocopying, recording or otherwise, except as permitted by the UK Copyright, Designs and Patents Act 1988, without the prior permission of the publisher.

Designations used by companies to distinguish their products are often claimed as trademarks. All brand names and product names used in this book are trade names, service marks, trademarks or registered trademarks of their respective owners. The publisher is not associated with any product or vendor mentioned in this book.

Limit of Liability/Disclaimer of Warranty: While the publisher and author(s) have used their best efforts in preparing this book, they make no representations or warranties with respect to the accuracy or completeness of the contents of this book and specifically disclaim any implied warranties of merchantability or fitness for a particular purpose. It is sold on the understanding that the publisher is not engaged in rendering professional services and neither the publisher nor the author shall be liable for damages arising herefrom. If professional advice or other expert assistance is required, the services of a competent professional should be sought.

Library of Congress Cataloging-in-Publication Data

Names: Chartered Institute of Building (Great Britain), author.
Title: Code of practice for programme management in the built environment / The Chartered Institute of Building.
Description: Chichester, UK ; Hoboken, NJ : John Wiley & Sons, 2016. | Includes bibliographical references and index.
Identifiers: LCCN 2015051191| ISBN 9781118717851 (pbk.) | ISBN 9781118717844 (epub)
Subjects: LCSH: Building–Superintendence–Great Britain. | Project management–Great Britain.
Classification: LCC TH438 .C46 2016 | DDC 658.4/04–dc23
LC record available at http://lccn.loc.gov/2015051191

A catalogue record for this book is available from the British Library.

Wiley also publishes its books in a variety of electronic formats. Some content that appears in print may not be available in electronic books.

Cover Image: Rawpixel Ltd/Getty

Set in 10/13pt Franklin Gothic by SPi Global, Pondicherry, India
Printed in Singapore by C.O.S. Printers Pte Ltd

1 2016

Contents

Foreword	ix
Acknowledgements	xi
List of Figures	xiii
Working Group (WG) of the *Code of Practice for Programme Management*	xv
Summary of Key Terminology	xvii

Introduction — 1
Building information modelling (BIM) and programme management — 4

CHAPTER 1

The Context of Programme Management — 5

1.1	Definitions of projects, programmes and portfolios	5
1.2	Understanding programme management: is there a programme?	10
1.3	Programme management in the built environment	10
	1.3.1 Corporate social responsibility (CSR)	12
	1.3.2 Sustainability and the environmental mandates	13
	1.3.3 Ethics in programmes: business and professional	13
	1.3.4 Health and safety standards and requirements	13
1.4	Types of programmes	13
1.5	Range and scope of programmes	14
1.6	Need for programme management	15
1.7	Programme management process and stages	15
1.8	Programme organisation structure	18
	1.8.1 Types of clients who may initiate programmes	18
	1.8.2 Client organisation structure	20
	1.8.3 Programme management structure	21
	1.8.4 Business partners	23
	1.8.5 Stakeholders	23
1.9	Portfolio management	25

CHAPTER 2

Stage A: Inception — 27
- 2.1 Purpose of stage — 27
- 2.2 Stage outline — 27
- 2.3 Stage organisation structure — 29
  - 2.3.1 Stage structure and relationships — 29
  - 2.3.2 Stage roles of key participants — 29
- 2.4 Programme management practices — 31
  - 2.4.1 Strategic change — 31
  - 2.4.2 Funding policy and strategy/arrangements — 35

CHAPTER 3

Stage B: Initiation — 37
- 3.1 Purpose of stage — 37
- 3.2 Stage outline — 37
- 3.3 Stage organisation structure — 39
  - 3.3.1 Stage structure and relationships — 39
  - 3.3.2 Stage roles of key participants — 39
- 3.4 Programme management practices — 42
  - 3.4.1 Benefits management — 42
  - 3.4.2 Feasibility study — 43
  - 3.4.3 Funding arrangements — 44

CHAPTER 4

Stage C: Definition — 47
- 4.1 Purpose of stage — 47
- 4.2 Stage outline — 47
- 4.3 Stage organisation structure — 52
  - 4.3.1 Stage overall structure and relationships — 52
  - 4.3.2 Stage roles of key participants — 52
  - 4.3.3 External environment and relationships: mapping the landscape — 57
- 4.4 Programme management practices — 58
  - 4.4.1 Scope management — 58
  - 4.4.2 Benefits management — 60
  - 4.4.3 Risk Management — 60
  - 4.4.4 Governance of programme management: steering for success — 64
  - 4.4.5 Issues management — 67
  - 4.4.6 Time scheduling — 67
  - 4.4.7 Financial management — 68
  - 4.4.8 Cost management — 69
  - 4.4.9 Change control — 72
  - 4.4.10 Information management — 72
  - 4.4.11 Communication/stakeholder management — 75
  - 4.4.12 Quality management — 77
  - 4.4.13 Procurement and commercial management — 78
  - 4.4.14 Health and safety management — 80
  - 4.4.15 Sustainability/environmental management — 80

CHAPTER 5

Stage D: Implementation — 83
- 5.1 Purpose of stage — 83
- 5.2 Stage outline — 83
- 5.3 Stage organisation structure — 84
  - 5.3.1 Stage structure and relationships — 84
  - 5.3.2 Stage roles of key participants — 85
- 5.4 Programme management practices — 89
  - 5.4.1 Performance monitoring, control and reporting — 90
  - 5.4.2 Risk and issue management — 90
  - 5.4.3 Financial management — 91
  - 5.4.4 Change management — 91
  - 5.4.5 Information management — 92
  - 5.4.6 Stakeholder/communications management — 92
  - 5.4.7 Quality management — 92
  - 5.4.8 Procurement and commercial management — 92
  - 5.4.9 Health and safety management — 93
  - 5.4.10 Sustainability/environmental management — 93
  - 5.4.11 Transition management – projects closure — 94

CHAPTER 6

Stage E: Benefits Review and Transition — 95
- 6.1 Purpose of stage — 95
- 6.2 Stage outline — 95
- 6.3 Stage organisation structure — 97
  - 6.3.1 Stage structure and relationships — 97
  - 6.3.2 Roles of key participants — 97
- 6.4 Programme management practices — 98
  - 6.4.1 Benefits management — 98
  - 6.4.2 Benefits and dis-benefits — 100
  - 6.4.3 Transition strategy and management — 107

CHAPTER 7

Stage F: Closure — 109
- 7.1 Purpose of stage — 109
- 7.2 Stage outline — 109
- 7.3 Stage organisation structure — 110
  - 7.3.1 Stage structure and relationships — 110
  - 7.3.2 Stage roles of key participants — 111
- 7.4 Programme management practices — 112
  - 7.4.1 Programme closure — 112

Appendices — 115
- T1 Vision Statement Template — 115
- T2 Programme Mandate Template — 117
- T3 Programme Brief Template — 118
- T4 Business Case Template — 121
- T5 Monthly Programme Report Template — 125

T6	Programme Highlight Report Template	126
T7	Benefits Profile Template	129
T8	Tracking Benefits: Benefits-Monitoring Template	130
T9	Programme Closure Report Template	131

Key Roles: Skills and Competencies   133
Programme Management Case Studies   145
    Case Study 1 – Example of a Vision-Led Programme:
        London Olympics 2012   145
    Case Study 2 – Example of an Emergent Programme:
        High Street Retail Store Re-branding   150
    Case Study 3 – Example of an Emergent Programme:
        Highways England   155

Bibliography   163
Index   165

Foreword

The concept of programme management is relatively new in the built environment. Its need, and continued growth, arises from the expectation that benefits obtained through coordinated management of multiple linked projects are greater than the sum of the individual project benefits. Therefore, programme management provides a systemic approach to achieve common goals and overall benefits.

Having started its life as a public sector tool, programme management has been gaining popularity in the private sector. There are now a significant number of organisations in both sectors, which are involved in the practice of programme management either in the capacity of client, programme manager, or both.

There are a number of documents and publications currently available for the general discipline of programme management. However, when it comes to the specific nature of the built environment – in which there are growing numbers of large and significant programmes – this new *Code of Practice* leads the way in being an authoritative document for both public and private sector practitioners.

Developed by representatives from the major professional institutions associated with construction and real estate, and from the key public sector organisations, practices and corporations involved with our industry, this document sets out best practice for programme management in the built environment.

I strongly commend the effort by this cross-institutional, public and private practice working group, in leading the way to produce this excellent *Code of Practice* for our industry. This will be of great value to all the associated clients, programme managers and supply chain professionals, as well as all students of the subject and their mentors. The benefits should be felt by not only those practising in the United Kingdom but also those globally, wherever programme management is gaining in importance as a delivery tool for programmes within the built environment.

Acknowledgements

This *Code of Practice for Programme Management* represents a continued effort over a sustained period of time, under the stewardship of Roger Waterhouse FCIOB and David Woolven FCIOB, to produce a practical document for a discipline which spans many industries. Programme management has no single universal definition or accepted standards, it aims to achieve benefits instead of just being time, cost and quality efficient; there are various types and a plethora of associated procedures and processes.

When the CIOB published the first edition of the *Code of Practice for Project Management* for construction and development in 1992, it was unique in many ways and has since found its place within our industry as an authoritative document. Its popularity has led to successive editions. Now into its fifth iteration, this pioneering document, I believe, will continue to serve the industry well.

This new *Code of Practice for Programme Management*, has similarly been prepared by a broad representation of the industry, with contributions from built environment specialists and interdisciplinary cooperation between professional institutions which represent our industry. I congratulate their perseverance and persistence in producing this excellent document and thank them all for their valued assistance in the process. A list of participants and the organisations represented is included in this book.

I would take this opportunity to extend a special note of thanks to Arnab Mukherjee FCIOB, for giving the document its final shape and coordinating the editing process.

Chris Blythe
Chief Executive
Chartered Institute of Building

List of Figures

0.1	Benefits cycle	2
0.2	Key output document at each stage	3
0.3	Key output document responsibility matrix	3
1.1	Establishing relatedness	7
1.2	Organisationally related projects	8
1.3	Key characteristics for projects, programmes and portfolios	9
1.4	Programme management in context	11
1.5	Programme delivery in built environment	12
1.6	Types of programmes	14
1.7	The programme's life	16
1.8	Programme organisation structure	19
1.9	Stakeholder map – illustrative example	24
1.10	Portfolio management structure	25
2.1	Stage A: Inception	28
2.2	Stage A: Inception – Organisation structure	29
2.3	Programme delivery in the built environment	32
2.4	Olympic Delivery Authority – London 2012	33
2.5	Strategic change and strategic objectives by change type	34
2.6	Strategic objectives alignment. HSSE – Health, Safety, Security & Environment	35
3.1	Stage B: Initiation	38
3.2	Stage B: Initiation – organisation structure	40
3.3	Benefit delivery in three stages	42
3.4	Benefits categories	42
3.5	Example of graphical representation of benefits realisation over time	43
4.1	Stage C: Definition	48
4.2	Contents of the programme delivery plan	52
4.3	Stage C: Definition – organisation structure	53
4.4	Stakeholder map	58
4.5	Three- point estimate triangle	62
4.6	Estimation of uncertainty: illustrative example	63
4.7	S-curve detailing the cumulative contingency requirement	63
4.8	Change management, risk management and reporting	65
4.9	Ability to impact and commitment to the change	66
4.10	Financial management roles and responsibilities	68
4.11	Programme budget for transport programme (example)	69
4.12	Delivery/project performance – programme EVM summary	70
4.13	Programme fiscal year performance (annual spend forecast)	71
4.14	Four-year programme cost projection	71
4.15	Reporting integration	72
4.16	Full year programme expenditure example	73

List of Figures

4.17	Invitation to tender (ITT) and signed outline contract (SOC) plus value of contract placed	80
5.1	Stage D: Implementation	84
5.2	Stage D: Implementation – organization structure	85
6.1	Stage E. Benefits review and transition	96
6.2	Stage E: Benefits review and transition – organisation structure	97
6.3	Managing and realising benefits	99
6.4a	Benefits map (leisure facility transformation programme): Step 1 – mapping programme objectives to strategic objectives	101
6.4b	Benefits map (leisure facility transformation programme): Step 2 – Identifying and mapping benefits to programme objectives	102
6.4c	Benefits map (leisure facility transformation programme): Step 3 – Identifying business changes	103
6.4d	Benefits map (leisure facility transformation programme): Step 4 – Mapping project outputs to benefits	104
6.4e	Benefits map (leisure facility transformation programme): Step 5 – Mapping the links between programme objectives, benefits, business changes and project outputs	105
6.5	Organisation size over time for programme delivery	108
7.1	Stage F: Closure	110

Working Group (WG) of the *Code of Practice for Programme Management*

Saleem Akram BEng (Civil) MSc (CM) PE FIE FAPM FIoD EurBE FCIOB	Director, Construction Innovation and Development, CIOB
Gildas André MBA MSc BSc (Hons) MAPM MCIOB	Managing Partner, GAN Advisory Services
David Haimes BSc (Hons) MSc MCIOB	Strategic Programme Director, Manchester Airports Group
Dr Tahir Hanif PhD MSc FCIOB FAPM FACostE FIC CMC FRICS	Project Control Specialist, Public Works Authority, (Ashghal), State of Qatar
Stan Hardwick FCIOB EurBE	Global Contracts Manager – Procurement, Specsavers
Dr Chung-Chin Kao ICIOB	Head of Innovation & Research, CIOB
Arnab Mukherjee BEng(Hons) MSc (CM) MBA FAPM FCIOB	WG Technical Editor
Andrew McSmythurs BSc FRICS MAPM	Director of Project Management at Sweett (UK) Ltd and RICS Representative
Paul Nash MSc FCIOB	Director, Turner & Townsend, Senior Vice President CIOB
Piotr Nowak MSc Eng ICIOB	WG Secretary, Development Manager, CIOB
Dave Phillips FAPM CEng	Divisional Director, Mott MacDonald
Milan Radosavljevic PhD UDIG MIZS-CEng	University of the West of Scotland
Dr Paul Sayer	Publisher, John Wiley & Sons Ltd, Oxford
Roger Waterhouse MSc FRICS FCIOB FAPM	WG Chair, University College of Estate Management, Royal Institution of Chartered Surveyors, Association for Project Management
David Woolven MSc FCIOB	WG Vice Chair/Editor – University College London

The following also contributed in development of the *Code of Practice for Programme Management*.

Susan Brown FCIOB MRICS	Property Asset Manager at City of Edinburgh Council
Jay Doshi	ICE Management Panel Member (Director, Amey Ventures)
Nikki Elgood	WG Administrator, CIOB
Una Mair	WG Administrator, CIOB
Simon Mathews	Director/HLG Associates
Gavin Maxwell-Hart BSc CEng FICE FIHT MCIArb FCIOB	Head of Contract Management, AREVA CIOB Trustee, Non-Executive Director, Systech International

Working Group (WG) of the Code of Practice for Programme Management

David Merefield	Head of Sustainability, Property, Sainsbury's Supermarket Ltd.
Alan Midgley	Medium Risk Reviewer, Cabinet Office Director, AGMidgley Ltd.
Dr Sarah Peace BA (Hons) MSc PhD	Consultant
David Philp MSc BSc FICE FRICS FCIOB FCInstES FGBC	Global BIM/MIC Director – AECOM, RICS Certified BIM Manager, CIOB Ambassador
Tony Turton MBA FICE	Project Development and Production Director, Highways England

Summary of Key Terminology

Benefits	A (directly or indirectly) measurable improvement resulting from an outcome perceived as an advantage by one or more stakeholders and that contributes towards one or more organisational strategic objective(s).
Benefits management	The identification, definition, monitoring, realisation and optimisation of benefits within and beyond a programme.
Benefits profile	Used to define each benefit (and dis-benefit) and provide a detailed understanding of what will be involved and how the benefit will be realised.
Benefits realisation manager (BRM)	Supports programme manager by taking the responsibility in benefits identification, mapping and realisation – ensures that necessary business benefits are realised.
Benefits realisation plan	Used to monitor realisation of benefits across the programme and set governing controls.
Business change manager (BCM)	Responsible for ensuring that the objectives have been sufficiently and accurately defined, managing the transition activities and undertaking and determining whether the intended benefits have been realised.
Business partner	Organisations that have a business or financial interest in the outcome of the programme.
Clients	Persons using the services of a professional entity or those who are procuring products or services from a professional entity. In legal context, a client may instruct a professional entity to act on the client's behalf. In the programme sense, this document defines clients as 'the body or group that procures the services of professionals to initiate and deliver projects or a programme of projects'.
Customer	Persons who are paying for a product or a service but not necessarily in the legal context represented by the professional entity.
Deliverable	What is to be provided as a result of an initiative or project – typically tangible and measurable.
Dis-benefit	A (directly or indirectly) measurable decline resulting from an outcome perceived as a negative by one or more stakeholders that may or may not affect one or more organisational strategic objective(s).
Issue	A relevant event that has happened or is likely to happen, wasn't planned and requires management action.
Opportunity	A relevant but uncertain event that can have a favourable impact on objectives or benefits.

Summary of Key Terminology

Outcome	The result of a change. Outcomes are desired when a change is conceived and are achieved as a result of the activities undertaken to reflect the change.
Output	The tangible or intangible effect of a planned activity or initiative.
Portfolio	A portfolio is a total collection of programmes and stand-alone projects managed by an organisation to achieve strategic objectives.
Programme	A programme is a collective of related projects coordinated to achieve desired benefits not possible from managing them as a group of individual projects.
Programme brief	Used to assess whether the programme is viable and achievable.
Programme communication manager (PrgCM)	Supports the programme manager by managing all internal and external communication channels, developing the programme communications plan and ensuring governance of internal and external communication protocols.
Programme delivery plan (PDP)	A detailed description of what the programme will deliver, how and when it will be achieved, financial implications of its delivery and implementation.
Programme financial manager (PrgFM)	Deals with complex financial issues including funding arrangements, cash flow and financial governance. Responsible for programme financial plan, budget and financial reporting.
Programme financial plan	A financial statement that collects all the costs that have been identified in relation to implementing the programme – often the funding streams are also identified in this document.
Programme management board (PrgMB)	A group established to support a programme sponsor in delivering a programme.
Programme management office (PMO)	The function providing the information and governance for a programme and its delivery objectives – it can provide support to more than one programme.
Programme manager (PrgM)	The role responsible for the setup, management and delivery of a programme – typically allocated to a single individual; for large and complex programmes an organisation can be given this role.
Programme mandate	Expansion of the vision statement setting out in greater detail what it is that the programme needs to achieve in terms of the outcomes and what it is that the programme seeks to deliver.
Programme monitor	In certain privately funded programmes, a programme monitor (sometimes known as funder/lender/investor's advisor or monitor) may be appointed, on behalf of the funding entities, to safeguard the interest of the funders.
Programme sponsor (PrgS)	The main driving force behind a programme and often is the point of accountability for the delivery.
Programme sponsor board (PrgSB)	The driving group behind the programme which provides the investment decision and senior level governance for the rationale and objectives of the programme.
Programme timescale plan	An overall delivery time schedule for the programme.
Project	A project is a temporary and transient undertaking created to achieve agreed objectives and produce and deliver a product, service or result
Risk	An uncertain event or set of events that, if it occurs, has an effect on the achievement of the objectives. A risk is measured by a combination of the probability of a perceived threat or opportunity occurring and the magnitude of its impact on objectives.

Stakeholder	Any individual, group or organisation that can affect, be affected by or perceives itself to be affected by a programme.
Transition	The changes that need to take place in business as usual, which are aimed to be managed, as project outputs are exploited in order to achieve programme outcomes.
Transition plan	The schedule of activities to cover the transition phase of the benefits realisation plan.
Vision	A view of a better future that will be delivered by the programme.
Vision statement	A business vision for change setting out the intent and the benefits sought.

The interpretations of the key terminologies are based on the current definitions and usage across a number of industries and current good practice. Some of the interpretations are specific to this document. Further references are includes in the Bibliography.

0 Introduction

This is the first edition of *Code of Practice for Programme Management in the Built Environment*. It is a natural development from the highly successful *Code of Practice for Project Management for Construction and Development*, now in its fifth edition, having been published initially in 1992. It, too, was the first *Code of Practice* for our industry for project management.

Both codes of practice were developed by representatives from the major professional Institutions associated with the built environment, the Chartered Institute of Building CIOB), Royal Institution of Chartered Surveyors (RICS), Royal Institute of British Architects (RIBA), Institution of Civil Engineers (ICE), Association for Project Management (APM) and from key government departments, industry practices and corporations, both domestic and international.

Just like project management, programme management is not unique to construction and real estate or the built environment, and there are many generic publications on programme management, not least those prepared by government. However, the term 'programme' has been used generically across many industries for decades, often in relation to extended projects or activity and time-related undertakings. We have researched many such publications on the way to achieving our aim of delivering a code of best practice for programme management for the built environment.

It is perhaps true to say that the earliest programmes of any strategic significance were those sponsored by government. Hence, much of the early research was focussed predominantly upon publicly funded programmes. This was not dissimilar to the situation for projects in the early days of project management, although privately funded projects were not far behind. However, today the term 'programme management' is still not fully understood by many professionals in the built environment. Many think of a programme as just a collection of projects. This *Code of Practice* is more specific and describes a programme as a collective of related projects coordinated to achieve desired benefits more effectively than when managing them as a group of individual projects.

Why did we choose 'for the built environment' and not 'for construction and development' as we did for the project management *Code of Practice*? Well, one of the key differences is that within the built environment there are many projects which are not construction or development related. For example, if we consider some of the client sectors involved in creating new facilities and/or infrastructure such as highways, rail, airports, shipping or nuclear and so on, all of these are likely to incorporate projects which are not related to construction. These may include disciplines such as information technology, human resources management (HRM), transportation, marketing and

Code of Practice for Programme Management in the Built Environment, First Edition. The Chartered Institute of Building.
© 2016 John Wiley & Sons, Ltd. Published 2016 by John Wiley & Sons, Ltd.

so on. Indeed, even mainstream construction developments may include similar disciplines as self-contained projects within their own programme.

There is another difference between 'project' and 'programme'. Whereas traditionally the former is measured by the criteria of time, cost and quality, the latter is determined and measured by the strategic objectives and the benefits to the client organisation which might otherwise not have been realisable had the projects been managed independently.

Furthermore, time has shown that where there are multiple (related) projects within a parent organisation which are being independently managed, issues such as the lack of coordination and overall control can arise. This can affect efficiency and effectiveness, which would lead to confusion over the responsibility and control of these projects.

From a stakeholder prospective, programmes can create value by improving the management of projects in isolation. This is particularly true where the working environment involves multiple small projects, where project integration, in terms of both development and deliverable benefits, is crucial for competitive success.

The need for programme management in the UK construction sector has probably arisen due to a number of factors: (i) The size of projects has generally increased in scope and expected benefits, and we expect more, sooner. (ii) Competition has increased. Just look at how the Olympics have developed from major projects to programmes, with each one trying to equal the previous country`s output/benefits. (iii) There is an increase in desired benefits due to growing ambition and availability of funding. Together, clients see a programme of related projects as posing less risk than a number of individual projects when considered alone.

The fifth edition for *Project Management* continues to provide the relevant guidance and procedural requirements for the successful management of individual projects. We have now developed the client-led single project into a group of related projects, the programme, for the corporate client. This *Code of Practice,* therefore, brings together the elements of functionality and procedures specific to the coordinated management and successful delivery of a number of related projects within the built environment focussed at the programme level.

The 'programme' is approached stage by stage, where the fundamental benefits cycle is identified (Figure 0.1). This is central to the drivers of the programme, with key functions and outputs (Figure 0.2) being highlighted in every stage and with clear ownership allocation for each output (Figure 0.3). Given the scope for variations in terminology and approaches that are possible within the practices of programme

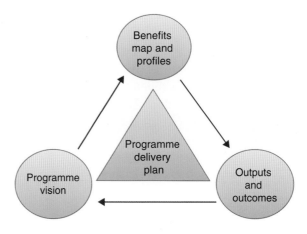

Figure 0.1 Benefits cycle.

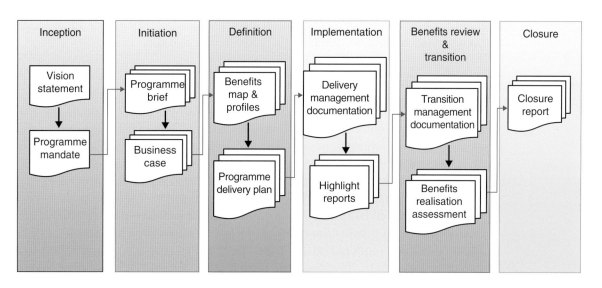

Figure 0.2 Key output document at each stage.

Document	PrgSB	PrgS	PrgMB	PrgM	BCM
Programme mandate	SO	A,D			
Programme brief	SO	A	R	D,C	D,C
Programme business case	SO	A	R	D,C	D,C
Programme definition plan	SO	A	R	D,C	D,C
Risks and issues register	I	A	SO	R,D	C
Stakeholder engagement and communications	I	A,SO	C	R,D	SO,C
Benefits realisation plan	I	A,SO	C	R,D	C,D
Programme delivery plan	I	A,SO	SO,C	R,D	C,D
Programme highlight report	I	A	SO	R,D	C
Programme transition plan	SO	SO	A,C	C	R,C,D
Programme outcomes review report	SO	A,SO	R,C	D	C

PrSB – Programme Sponsor Board PrgS – Programme Sponsor
PrgMB – Programme Management Board PrgM – Programme Manager
BCM – Business Change Manager

SO – Sign Off: the person or group with the authority to accept that the document satisfactorily addresses the key question and can commission the next step of work

A – Accountable: the person or group who is held accountable for the quality of the document and for deciding that it is fit for purpose

R – Responsible: the person or group who is responsible for getting the document produced on behalf of the accountable person or group

D – Doer: the person or group who does the work to complete the document

C – Consulted: the person or group who needs to provide input to the document to ensure that the content is fit for purpose

I – Informed: the person or group who needs to be informed of the content of the document

Figure 0.3 Key output document responsibility matrix.

management, this code establishes a clear and consistent process for managing programmes, regardless of their size, nature or complexity.

Building information modelling (BIM) and programme management

BIM is a collaborative way of working, underpinned by the digital technologies which unlock more efficient methods of designing, delivering and maintaining physical built assets. BIM both embeds and links key product and asset data in a series of domain and collaborative federated models, consisting of both 3D geometrical and non-graphical data. These models are prepared by different disciplines during the project life cycle within the context of a common data environment. The project participants provide defined, validated outputs via digital data transactions using proprietary information exchanges between various systems in a structured and reusable form. These digital computer models can be used for effective management of information throughout an asset's life cycle, from earliest concept through to operation. BIM has been described as a game-changing information and communications technology (ICT) and cultural process for the construction sector. A number of countries globally are starting to realise the opportunities BIM brings and are now investing heavily to develop their own capability. BIM processes are now mainstream to both new buildings and infrastructure and have great value in retrofit and refurbishment projects where complimentary technologies such as laser survey techniques and rapid energy analysis are employed.

At the heart of successful programme management is communication, information exchange and integration. BIM, as a management tool, has the ability to influence successful programme management, particularly in context of programmes involving capital assets.

The UK Government's Digital Built Britain (DBB[1]) strategy, built around the comprehensive utilisation of BIM tools, asset telemetry and real-time performance data, uses advanced computer systems to build 3D models of infrastructure and hold large amounts of information about its design, operation and current condition. At the planning stage, it enables designers, owners and users to work together in an integrated and concurrent manner to produce the best possible designs and to test them in the computer before they are built. In construction it enables engineers, contractors and suppliers to integrate complex components, cutting out waste and reducing the risk of errors. In operation it provides customers with real-time information about available services and maintainers with accurate assessments of the condition of assets.

In context of programme management, BIM can prove to be a vital management tool, utilised both at project level and at the programme level, to become a critical and necessary part of the data and information exchange, as well as the delivery of the programme and the benefits envisaged.

The decision to utilise BIM may be taken at the initiation stage and should form part of the programme business case.

---

[1] Digital Built Britain, (2015) *Level 3 Building Information Modelling-Strategic Plan*. HM Government. (available at www.bis.gov.uk).

1 The Context of Programme Management

- What are projects, programmes and portfolios?
- Why is there a need for programme management in the built environment?
- What are the contextual issues for programme management in built environment?
- What are the types of programmes?
- What are the stages of programme management?
- How is a programme organisation structured?
- What is the importance of stakeholders in programme management?

1.1 Definitions of projects, programmes and portfolios

Project

Projects are needed in every industrial sector, and several definitions of the term 'project' exist today. Some of the most commonly used definitions are listed below:

> A unique set of co-ordinated activities, with definite starting and finishing points, undertaken by an individual or organization to meet specific objectives within defined schedule, cost and performance parameters. ISO 21500: 2012/BS 6079 – 1:2010

> A project is a time and cost constrained operation to realize a set of defined deliverables (the scope to fulfil the project's objectives) up to quality standards and requirements. International Project Management Association (IPMA)[1]

> A unique, transient endeavour undertaken to achieve planned objectives. Association for Project Management (APM)[2]

---

[1] Quoted from IPMA Competence Baseline, Version 3.0 (ICB 3.0), 2006. Available from http://www.ipma.world/resources/ipma-publications/ipma-competence-baseline/. Accessed January 2016.
[2] See https://www.apm.org.uk/glossary.

Code of Practice for Programme Management in the Built Environment, First Edition. The Chartered Institute of Building.
© 2016 John Wiley & Sons, Ltd. Published 2016 by John Wiley & Sons, Ltd.

> A temporary endeavour undertaken to create a unique product, service or result. Project Management Institute (PMI)[3]

These definitions collectively recognise *temporary and transient nature* as the two fundamental characteristic of a project. Projects are temporary in that they have a definitive start and an end. They are also transient because they are completed as the organisation moves from one project to another at a different location and so on. Projects are created to *achieve agreed objectives and produce and deliver a product, service or result*. The involved parties need to agree to the objectives, and the partner tasked with achieving the objectives needs to first produce and finally deliver what has been set in the objectives.

The *CIOB Code of Practice for Project Management* (fifth ed., p. 317) defines project as:

> A unique process, consisting of a set of co-ordinated and controlled activities with start and finish dates, undertaken to achieve an objective conforming to specific requirements, including constraints of time, cost and resources.

The task of project management is to bring in at the right time and co-ordinate many different professionals and specialists to enable them to achieve the agreed objectives. To do this effectively, project managers need to manage key business functions for a project.

Programme

Programme is different from a project, but the two terms are often used interchangeably. Some of the notable existing definitions recognise the following:

> "A programme is designed as a temporary flexible organisation structure created to coordinate, direct and oversee the implementation of a set of related projects and activities in order to deliver outcomes and benefits related to the organisation's strategic objectives." Business Innovation and Skills (BIS)

> "A program is a group of related projects managed in a coordinated manner to obtain benefits and control NOT available from managing them individually. Programs may include elements of related work outside of the scope of the discreet projects in the program… Some projects within a program can deliver useful incremental benefits to the organization before the program itself has completed." PMI

> "A group of related projects and change management activities that together achieve beneficial change for an organisation." APM

Programmes comprise multiple related projects, and that by itself makes a programme distinctly different from a project. Programmes are often ongoing, with a number of milestones, and do not necessarily have the strictly finite nature of a project. Even when a programme has an end date, the time scheduled is normally far longer than any project within this programme. Unlike projects, programmes are created for horizontal co-ordination of projects, which often run in parallel.

---

[3] Definition available at the PMI web site, What Is Project Management? See http://www.pmi.org/About-Us/About-Us-What-is-Project-Management.aspx (accessed January 2016).

From a business and customer perspective, a programme is a temporary organisation designed to operate, learn and adapt in a complex environment of interrelated projects, people and organisations. In this context, the programme manager is the chief executive officer of a temporary organisation with the ability to carry the flame for what users want.

A programme therefore comprises *a collective of related projects* which are limited in time and designed to individually deliver agreed upon objectives and which produce and deliver a product, service or result. The coordinated manner by which they are managed *delivers programme benefits* that are greater than the sum of individual project benefits were they not coordinated at the programme level. Success of a programme is thus dependent on a programme team's ability to deliver those benefits.

In the context of construction, CIOB defines a programme in the following way:

> A programme is a collective of related projects coordinated to achieve desired benefits more effectively from managing them as a group of individual projects.

From a summary point of view, a business or a client define strategic objectives that are implemented through a programme of interrelated projects internally delivered or outsourced to specialist supplier and/or contractors.

Projects in a programme may be related in different ways. For example, a number of projects that collectively need to be managed to deliver a set of benefits in order to address client's objectives are related through client or customer/end user (see Figure 1.1).

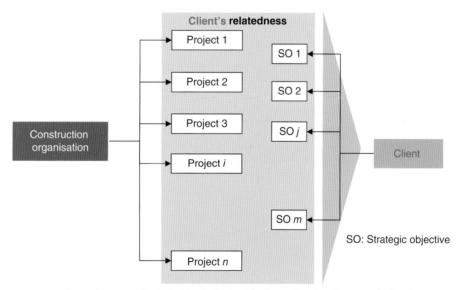

Programme formed as a result of client's strategic objectives. Construction organisation is may need to adjust their processes to reflect the *client's* requirements across all of their projects.

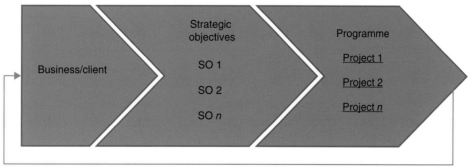

Figure 1.1 Establishing relatedness.

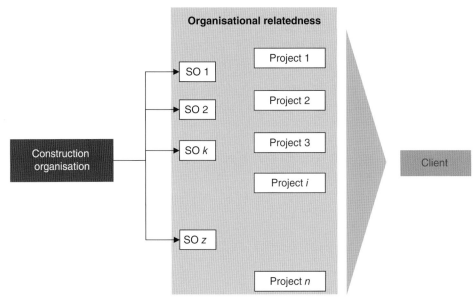

Programme formed as a result of changed or new strategic objectives that require a concerted effort through a number of projects to deliver the desired benefits for the organisation and *clients*.

Figure 1.2 Organisationally related projects.

On the other hand, a number of projects that collectively need to be managed to deliver a set of benefits in order to address construction organisation's strategic objectives are organisationally related (see Figure 1.2).

The task of programme management is to create and co-ordinate a collective of related projects in order to deliver programme benefits which would not be as achievable if they are managed as a group of individual projects.

Portfolio

The existing definitions of 'portfolio' recognise that organisations are involved in a number of programmes and projects at any given time that may or may not be related.

> "The term Portfolio is used to describe the total set of programmes and stand-alone projects undertaken by an organisation." BIS

> "A [portfolio is a] grouping of an organisation's projects, programmes. Portfolios can be managed at an organisational or functional level." APM

> "A portfolio is a collection of programs, projects and/or operations managed as a group. The components of a portfolio may not necessarily be interdependent or even related – but they are managed together as a group to achieve strategic objectives." PMI

However, a portfolio is not a random collection. Organisations need to achieve their strategic objectives, so they carefully consider the kind of projects and programmes that constitute their portfolios. CIOB defines portfolio in the following way:

> A portfolio is a total collection of programmes and stand-alone projects managed by an organisation to achieve strategic objectives.

The task of portfolio management is to manage and maintain all of an organisation's projects and programmes to help achieve its strategic objectives. The organisation's ongoing business may be project-based and/or require projects and programmes to achieve the desired change to sustain its business.

Figure 1.3 summarizes the key characteristics of project management, programme management and portfolio management as set out above.

	Project management	**Programme management**	**Portfolio management**
Definition	Management of endeavours (projects) which are temporary and transient and deliver an output, product, service or result	Management of a collective of related projects (programme) coordinated to achieve desired benefits which will not be possible if the projects are managed as a group of individual projects	Management of the total collection of programmes and projects to achieve strategic objectives
Scope	Relatively narrow, focused on delivery of defined results/products/services/outputs	Relatively wider, focussed on capabilities that will enable achievement of the desired benefits	Scope includes those of all the programmes and projects within the portfolio.
Change	Change is rigidly controlled to minimise impact on time, cost and scope	Change is inevitable and has to be embraced within the procedure by continuous review of the business case.	Change has to be risk evaluated and optimised to achieve the desired objectives for which the investment decision has been made
Schedule	Detailed activity-based schedule is typically prepared to manage delivery of defined outputs	High-level scheduling to provide an overview of components and to establish priorities, interdependencies and conflicts	Overall process and reporting to establish costs and contributions
Duration	Relatively shorter time, typically expressed in weeks or months, needed to create and deliver the defined outputs	Relatively longer time, typically expressed in years, needed to create the new capability and achieve the desired benefits	Ongoing activity with no anticipated end date; the inherent part of day-to-day running of the organisation
Governance	Management of monitoring and control of tasks and activities to ensure budget, time and specification parameters are met	Management of component projects and overview of the programme as a whole to ensure business case parameters are met	Management and overview of the whole portfolio to ensure organisational objectives are met
Success	Delivery of the output within the budget, time and specification criteria and to the satisfaction of the users	Achievement of benefits desired	Achievement of organisational objectives

Figure 1.3 Key characteristics for projects, programmes and portfolios.

1.2 Understanding programme management: is there a programme?

When is it a project?

An undertaking is considered and executed as a project when:

- the delivery criteria, scope, quality, cost and time can be defined and measured
- the delivery structure and methodology is known and available

The output/benefits resulting from the project may or may not deliver the total outcome required by the undertaking's initiator.

When is it a programme?

An undertaking is considered and executed as a programme when:

- the delivery criteria may or may not be fully known, defined or approved
- the undertaking requires a high level of regulated governance
- achievement of the overall outcome required necessitates a number of related projects, each demanding different specialist skills, expertise or organisational approaches
- the size, complexity and uncertainty of the undertaking are such that delivery is best approached by creating a number of projects
- the delivery skills required are beyond the organisational and contractual arrangements for one team
- the geographic spread of the undertaking makes it uneconomic or infeasible to have one project
- time or cost constraints mean that it is uneconomic or infeasible to have one project
- there is a requirement to manage interdependencies between projects
- there is a requirement to manage conflicting priorities and resources across projects

When is it a portfolio?

An undertaking is considered and executed as a portfolio when:

- the programmes and projects comprising the undertaking are discrete, with no related or shared output
- there is a requirement to manage priorities and resources across programmes and projects
- there is a requirement for high-level governance and monitoring of the programmes and projects

A portfolio can relate to an organisation's total activity or to specific undertakings

1.3 Programme management in the built environment

We have defined the terminology for projects, programme and portfolio, change and programme management practices in the context of a business organisation. We now consider these in the context of the built environment.

Change management addresses introducing new things or doing things better for continuous improvement: delivering organisational changes (building new or improving existing assets) while managing quality and risk, as well as producing measurable and sustainable benefits adding value to operating business models.

Programme management brings reality and method to strategic planning and to the delivery of strategic change with the right strategic objectives phased and delivered on cost, on time, on quality and on benefits.

Project management is deterministic, learning rules and applying them well, and delivering output and products or single assets.

Portfolio management encompasses programme or projects that are not necessarily linked, collectively delivering a strategy.

Programme management in this context is illustrated in Figure 1.4 in a circular, continuous process running from strategy (top left corner) to operations and benefits realisation (bottom left corner). It is contained between business as usual, strategizing and realising benefits while operating a business or an organisation and (right-hand side) planning and managing a programme to deliver outcomes and benefits (see Figure 1.5).

In the context of the definitions as outlined in this document, many major projects are actually programmes comprising several related projects, but they have not been treated as programmes. The construction sector has gradually shifted the focus from products to services, where a product is one of the deliverables. Now, programmes and their associated projects don't just deliver buildings, they deliver a product, benefit or outcome which adds value their customers seek (i.e. 'servitization[4]'). Such an

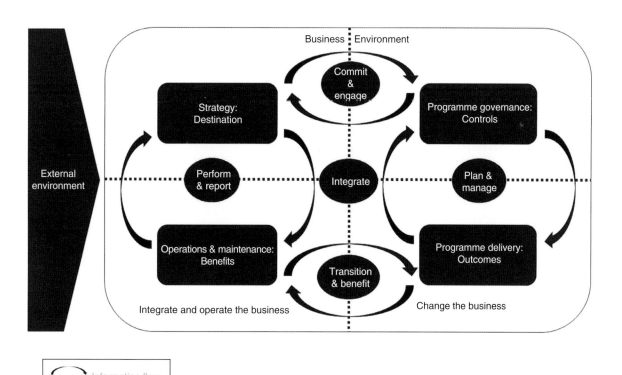

Figure 1.4 Programme management in context.

---

[4] Servitization is the process by which a manufacturer changes its business model to provide a holistic solution to the customer, helping the customer to improve its competitiveness, rather than just engaging in a single transaction through the sale of a physical product.

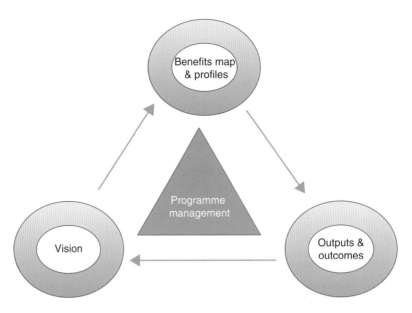

Figure 1.5 Programme delivery in built environment.

outcome calls for a holistic approach that may well require several very different, yet related projects delivered for the same client (e.g. several building projects, infrastructure projects, training projects, IT projects, etc.).

The London Olympics 2012 represents a good example of a programme where individual venues, infrastructure, legacy and so on each represented separate sets of related projects all under the umbrella of the Olympic Delivery Authority (ODA). The ODA were tasked to act as programme manager to achieve the desired benefits that would not have been achieved had they been managed individually.

The traditional linear delivery cycle for projects offers a robust model for programme planning and delivery. Using this model, an effective programme will anticipate and integrate organisational changes within financial, contractual and physical constraints as time passes. Innovation through clarity of strategic intent and creative design coupled with adapted controls (risk management and change management) is critical to successful scope delivery (on time, to budget and quality) and benefits realisation. The programme and organisational structure (including resource levels) will be gate controlled and will evolve through each delivery phase.

1.3.1 Corporate social responsibility (CSR)

Programmes and their associated projects need to embrace all the aspects normally associated with CSR and preferably some that go beyond 'responsibility' and 'compliance' and into the realm of voluntary environmental and social benefit.

Ideally, this will develop into a corporate business model, with self-regulation to ensure active compliance with both the literal meaning, and spirit, of the law.

CSR should consider and provide benefits in sustainable development, including an agenda which continually reduces the carbon footprint for each of the associated projects, where appropriate. It should begin by embracing all legislation associated with health and safety, environmental assessment, energy, waste and pollution, then progress through regular monitoring to achieve the maximum benefit for users, the environment, and the client organisation and its stakeholders.

CSR encompasses a number of disciplines associated with programme management, namely, sustainability, ethics, and health and safety, as well as the promotion of positive social and environmental change.

1.3.2 Sustainability and the environmental mandates

Environmental performance and impact, together with the other sustainability elements of 'economic' and 'social', may be particularly important to the client.

Environmental mandates include requirements for the environmental performances within related projects which comprise the programme. This includes all requirements on carbon emissions and energy consumption.

In addition, it may also prescribe requirements for the environmental impact on local topographies or areas adjacent to related projects. It may determine outcomes in terms of associated communities, such as providing employment and training opportunities or the use of supply chains.

Finally, the environmental mandates should also outline the overall environmental management criteria for related projects, including the key success factors for the related projects in terms of environmental management.

1.3.3 Ethics in programmes: business and professional

Programmes encompass a wide range of organisations and individuals, which may be divided into business and/or professional relationships. These parties vary in their practices, procedures, negotiations, management styles and financial and business processes. But all need to have ethical standards of trust and behaviour which are mutually acceptable, even if they contain variations, if their relationships within the programme are to be successful and sustainable.

Business and professional ethics need to comply with the local and business legal and statutory requirements. These should conform to the respective government legislation along with accepted international practices and relevant national and international professional standards.

1.3.4 Health and safety standards and requirements

Generally, the UK laws governing health and safety relate to all construction, manufacturing and engineering and are not industry specific. Good practice should be adhered to throughout all projects and programmes, irrespective of the location of the projects or programmes.

For further details see Briefing Note 3.01 (the CIOB code of Practice for Project Management, fifth ed. 2014, p.113).

1.4 Types of programmes

In the context of the built environment, programmes may be categorised in three key types, as illustrated in Figure 1.6, depending on the driver for change. These are (i) vision-led programmes which are necessarily top driven and set out to meet a particular strategic vision or need, (ii) emergent programmes, where the organisation recognises that a group of existing projects would be better managed together (a bottom-up approach) and (c) compliance programmes which are created in response to internal or external stimuli, often generated outside the control of the organisation.

All programmes, regardless of the driver and type, introduce change: internal or external or even both internal and external.

Example 1: Vision-led programmes

Reducing the carbon footprint of existing and new hospitals is a vision which requires a number of diverse retrofit and new-built projects according to commonly accepted

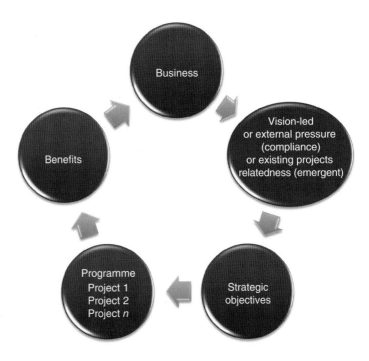

Figure 1.6 Types of programmes.

standards, staff training, logistics and other projects. Mobilisation of vast resources across geographical regions calls for co-ordination far beyond the needs of a single project.

Example 2: Emergent programmes

An organisation facing similar problems in several existing projects recognises that they need to be addressed across all projects. For example, some common factors that lead to delays and other types of losses need to be addressed holistically and solutions developed and implemented in all projects. A programme could then be formed by pulling together existing projects to develop and implement solutions across all projects in the programme.

Example 3: Compliance programmes

A change in facilities-related legislation could force an organisation with a large number of facilities to form a programme to implement the necessary changes across all facilities. Large organisations where compliance-related changes normally lead to the formation of programmes designed to implement the necessary changes normally need to maintain in-house programme management capacity.

1.5 Range and scope of programmes

Programmes can be of different sizes and complexity. Some will a number of dozens of projects over many years or continuous programmes with major milestones. Other programmes will be fairly small with only a few of projects, but regardless of the size, all programmes should achieve benefits set by the strategic objectives.

Programme scope should thus clearly spell out the anticipated benefits for using a programme mandate and or a programme brief and how a collective of projects aims to achieve those benefits to accomplish the strategic objectives. The benefits realisation process is thus a prerequisite for setting strategic objectives and in its initial phase defines the required projects to achieve the desired benefits.

1.6 Need for programme management

The need for programme management arises from the expectation that programme benefits obtained in a coordinated manner are usually greater than the sum of individual project benefits obtained in isolation. Programme management looks at overall benefits (i.e. programme optima) and helps individual project managers running their projects relative to other projects in the programme in terms of their progress, risks, resource requirements and sharing, knowledge sharing, issues, constraints and so on (see Section 4.3).

If projects are managed in an uncoordinated manner, individual project managers may use different tools and techniques, rendering any meaningful comparison of progress impossible. In addition, any resource and knowledge sharing would be left to chance or the willingness of individual project managers to share. In this way, programme management is a systematic approach to project grouping for the purpose of achieving common goals and overall benefits without leaving the process to chance.

1.7 Programme management process and stages

Whether in mature economies or in emerging markets, countries and organisations are looking for creative solutions to achieve their long-term vision and fulfil their citizens' and customers' needs. Government and business managers have to find ways to be both strategic and more efficient in developing and directing funding and managing strategic change programmes, in order to deliver their intended benefits.

The speed, scale and disruptive nature of change in the global economy, acceleration of knowledge distribution and technological improvement have led organisations to undertake more risky and unique initiatives into unknown territories.

Managers and their organisations are required to introduce strategic change in order to be more competitive, improve profitability, reduce effort, improve working environment, deliver better products, stimulate teamwork or achieve higher client satisfaction. They seek to improve an aspect of production or quality, to reduce delays or to build new assets or whatever management thinks will improve the triple bottom line.[5] And change is never easy, so managing change effectively within an organisation is vital to its success.

This section describes the approach for effective programme delivery using examples.

The following stages, illustrated in Figure 1.7, determine a framework for through-life managing of programmes:

- Stage A: Programme inception
- Stage B: Programme initiation
- Stage C: Programme definition
- Stage D: Programme implementation
- Stage E: Benefits review and transition
- Stage F: Programme closure

---

[5] The accounting framework of three elements relating to measurement of organisational performance in the context of people (social), planet (environmental) and profit (financial) originated from the concept of sustainable development.

1 The Context of Programme Management

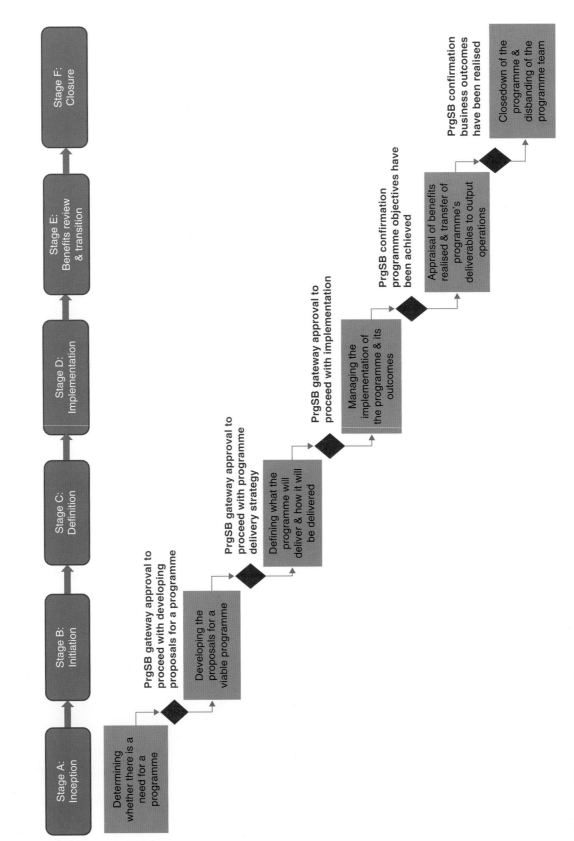

Figure 1.7 The programme's life.

Stage A: Programme inception

It is the programme sponsor's responsibility to outline programme vision, objectives and benefits. Vision statement and programme mandate are the two main outputs at this stage. However, optimum benefits may require a much more detailed benefits identification process managed by a business change manager and/or a benefits manager. There should be a clear need for a programme, and this need must arise from the desired benefits that have been previously identified (i.e. benefits register). This iterative process is important, particularly as benefits may be accompanied by dis-benefits, unintended benefits and consequential benefits that may not be immediately obvious.

Stage B: Programme initiation

A programme manager may be appointed at this stage to work along with the benefits manager in order to achieve the desired benefits. A programme manager develops a programme brief and, through identified benefits, produces a detailed business case, which is either accepted and the programme is given a go-ahead or rejected and potentially sent for review. Once the programme is accepted, the programme manager establishes governance mechanisms and develops the relevant plans.

Stage C: Programme definition

In some cases, a programme manager is appointed at this stage, but the overall goal at this stage is to have a fully functional programme management office (PMO) to develop the programme delivery plan (PDP). Identified benefits should be reviewed at this stage and detailed benefit profiles developed in order to produce the benefits realisation plan. This is accompanied by the projects register to enable programme-wide monitoring for the purpose of achieving programme outcomes.

Stage D: Programme implementation

Each project in a programme makes a contribution towards achieving the overall anticipated benefits, but these may be accompanied by dis-benefits and consequential benefits so it is important that their contribution is implemented and reviewed at regular intervals. While some benefits will be immediately realisable, most will only be fully achieved once the programme has closed or is undergoing a complete cycle review. Programmes may be very long, so these regular review points can be used to make necessary alterations in forthcoming projects in case there is an indication that a planned course may not produce the desired outcomes and benefits. In addition, this stage monitors and controls the established programme across all of its constituent projects. This includes appointment of project managers, performance of individual projects, interfaces between projects, benefits realisation, financial expenditure and programme changes.

Stage E: Benefits review and transition

In addition to regular review points, continuous programmes need major reviews after the anticipated benefits have been fully achieved to address the changes in strategic objectives or any external changes that lead to further projects and continuation of the programme. If a programme is not to be resumed later, then there will be a point when it needs to be closed. The closure of a programme needs to follow the programme outcomes review processes to establish whether the programme has delivered the outcomes as set out during stages A and B. Programme closure must demonstrate that benefits have been achieved and projects are completed successfully, review the performance of the programme, identify best practice and lessons for future programmes, update the benefits register and ensure ongoing benefits realisation after the programme has closed.

Stage F: Programme closure

At the point at which the programme sponsor has agreed that the required outputs and the outcome defined in the PDP has been achieved and no further works are envisaged, the programme is considered complete. The stage involves the controlled shutting down of the programme activities and the disbanding of the programme team.

1.8 Programme organisation structure

As described in Section 1.3, in general programmes may be vision led, emergent or compliance based (driven by an external pressure). That means some programmes will be internal, but others will be developed on behalf of an external client. Regardless of the case, a programme is always developed to deliver benefits to a client (internal or external).

Figure 1.8 shows an example of a programme organisation structure.

In terms of the nature of business, approaches and structure, the private sector is far more diverse than the public sector. However, it is common that interim or final outcomes and deliverables will need to be reported to a board of directors.

1.8.1 Types of clients who may initiate programmes

Clients are 'person (s) or organisation(s) using the services of a ... professional person or organisation'[6] or 'organizations or individuals for whom a construction project is carried out'[7]. In the context of the built environment, they may be defined as the body or group that procures the services of professionals or a customer who buys goods or services from a supplier or business.

In the programme sense, this document defines clients as the body or group that procures the services of professionals to initiate and deliver projects or a programme of projects. It is the clients who represent their organisation or the stakeholders, such as shareholders, who will engage professionals to initiate a programme of projects. These projects may indeed include non-construction projects, for example, IT, HR (human resources), and so on, as well as construction projects.

The client and the client's organisation will usually 'own' the programme and the projects, which may be from the public or private sector and likely to have obtained their own funding. The 'sponsor' may hold a similar position, but in construction development the sponsor is the person who normally represents the client and is usually in charge of the funding arrangements as well as serving as the main representative.

Programmes vary considerably in size and form in the same way that projects do. However, as programmes include multiple related projects, they have a tendency to be larger, be at a more 'strategic' level and take place over a longer time period than individual projects. Programme clients therefore are often at an established senior level in industry or government and may have progressed from being a client for a single project.

Alternatively, the programme may have started as just one or two related projects, or even a small group of projects, which has then enlarged to encompass even more projects over a longer time period. This extension could be caused by a variety of factors, such as new ownership or leadership, revised shareholder or stakeholder

---

[6] *Oxford Dictionary* See http://www.oxforddictionaries.com/definition/english/client (accessed January 2016).

[7] Health and Safety Executive. (2015) *Managing Health and Safety in Construction: Construction (Design and Management) Regulations 2015*. Available at http://www.hse.gov.uk/pubns/priced/l153.pdf.

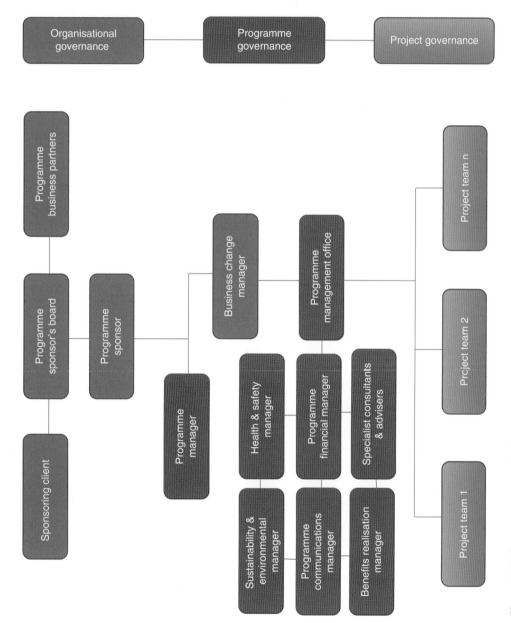

Figure 1.8 Programme organisation structure.

policy, new merger or acquisitions, planned or unplanned change in corporate strategy, increase in funding availability, etc.

The following lists typical organisations that may be public or private programme clients or sponsors, each having its own representative who could be termed the client, sponsor or programme manager. All at some point may have previously acted as a project client.

Typical public clients:

- Central government
- Local government
- Government departments
    - Health
    - Education
    - Heritage
    - Transport
    - Energy

Typical private clients:

- Banking and finance
- Retail
- Hospitality and leisure
- Manufacturing
- Food and beverage outlets
- Developers

1.8.2 Client organisation structure

Whether controlling single or multiple programmes, clients will allocate a number of roles overseeing the delivery of the programme. These may include the following groups or individuals:

Programme sponsor

A programme sponsor (PrgS) outlines programme vision, objectives and benefits. Directly responsible for developing the vision statement into the programme mandate, a PrgS's key responsibilities include:

- strategic direction and fit with the overall business strategy
- releasing the required resources
- ensuring programme stability in terms of time, budget and scope
- championing the programme at the most senior level
- providing high-level feedback on programme progress

The role of programme sponsor is a very senior position requiring visionary capabilities and competent leadership skills acquired from leading diverse senior teams.

In some organisations the programme sponsor role may be known as the senior responsible owner or the programme director.

Programme sponsor's board

The programme sponsor's board (PrgSB) will have the authority to make key decisions and commit expenditure on the programme on behalf of the sponsoring organisation. It will consist of executive-level individuals who are heads of functions of either the sponsoring organisation or of the final commissioned enterprise. Membership of the PrgSB needs to be ratified at the highest level, by the executive board or chief executive officer of the sponsoring organisation.

Throughout the programme the role of the PrgSB is to provide the overall strategic direction, support the programme sponsor in the implementation of the programme, ensure that adequate resources are available to the programme, monitor the programme's progress towards achieving the required outputs, facilitate the resolution of any major issues, determine when the programme's objectives have been achieved and ratify closure of the programme.

The PrgSB may also be known as the programme steering committee.

1.8.3 Programme management structure

The complexity of delivering a programme requires a team of highly experienced and skilled managers, which will include the following individuals or groups.

Programme manager

The programme manager (PrgM) coherently manages programme stages, reports to the programme sponsor and is responsible for the delivery of the proposed change. Key responsibilities include:

- developing and maintaining a project-supportive programme environment
- working with PrgS to ensure the programme is delivered on time, within budget and scope
- managing the PMO
- delivering programmes successfully in terms of agreed objectives and identified benefits (i.e. programme finances and benefits)

Programme manager is a senior appointment. The person in that role must have the necessary skills to implement the programme. In addition to core project management skills, a programme manager should have:

- sound understanding of business case development
- good knowledge of key programme-level financial and business indicators
- senior-level credibility to effectively support project teams
- excellent stakeholder management skills

Programme management board

The programme management board (PrgMB) is composed of senior managers of the programme management structure and provides advice and support to the programme manager. Its key responsibilities include:

- reviewing progress
- highlighting and resolving any issues that may be hindering progress

Programme management office

The PMO is a central support unit that supports the programme manager in overseeing day-to-day operation of a programme. A PMO includes a PMO manager and other specialist staff to carry out the functions required to:

- develop the programme delivery plan
- develop and maintain standards
- establish the governance controls
- manage programme documentation
- enhance capability to deliver the programme

Business change manager

Whereas the principal expertise of the PrgM is in managing the successful delivery of the programme in accordance with the required outcomes, the role of the business change manager (BCM) is much more focused on ensuring that the programme delivers the necessary business benefits. The BCM serves as the linchpin for the sponsoring organisation, the programme and the finished undertaking by ensuring the objectives have been sufficiently and accurately defined by managing the transition arrangements between the programme and the undertaking and by determining whether the intended benefits of the programme have been realised. The depth of knowledge and understanding required of the sponsoring organisation, and the way it operates, together with what is required from the new undertaking and the impact this will have on the existing operations, means that the BCM is likely to be a senior manager from within the sponsoring organisation.

In some organisations the BCM function may be called the business case manager.

Benefits realisation manager and/or business change manager

The programme benefits realisation manager (BRM) supports the PrgM by taking the responsibility for benefits identification, mapping and realisation. In effect, the BRM ensures that the programme delivers the necessary business benefits. The role of the BRM is to develop benefit profiles, assemble the benefits realisation plan, carry out benefit reviews and measure the effectiveness of the benefits.

Programme financial manager

The programme financial manager (PrgFM) deals with complex issues around tax liability, capital allowances and programme funding and is responsible for the programme financial plan, programme budget and financial reporting.

Considering the complexity of programmes at the interface between programme and business environments, the PrgFM should have sufficient experience with financial accounts and reporting in a project-based business environment.

Programme communications manager

The programme communications manager (PrgCM) supports the programme manager by managing all internal and external communication channels. The PrgCM develops the programme communications plan and must have sufficient experience with public relations as well as internal communication protocols within a project-based environment.

Programme monitor

In certain privately funded programmes, a programme monitor (sometimes known as funder/lender/investor's advisor or monitor) may be appointed on behalf of the funding entities to safeguard the interest of the funders.

The scope of services for the role of a programme monitor will depend on the individual programme(s) and its stage in the programme lifecycle.

Functions of programme monitor may include:

- representing the interests of the funding business partner(s)
- attending with the funder monitoring meetings between the programme commissioner and programme's delivery partners at agreed upon intervals
- monitoring the programme risks, in particular to those influencing funding, cash flow and potential income streams, and report back to the funder

1.8.4 Business partners

Programmes are likely to consist of a collection of organisations that have an active interest in the undertaking. These interests may be a shared ownership of the programme, shared governance, a financial sponsoring interest or an operational interest.

1.8.5 Stakeholders

Stakeholders include persons and organisations that have an interest in the strategy of the organisation and programme, have an impact or are impacted by a programme.

Programmes and organisations have to identify all stakeholders and assess the level of power they hold to affect the decisions and outcomes of the programme.

Stakeholder can be divided in two groups:

- Internal stakeholders: members of the organisations and those with an economic or contractual relationship with the programme
- External stakeholders: those with interest in the organisation and programme activities or those impacted by the activities in some way, such as governments, the public, interest and pressure groups, media and news organisations, local communities and statutory authorities

A list of common stakeholders may include the following:

- general public (people who are only indirectly effected by a programme but who may have a significant influence on its realisation)
- community (people who are directly effected by a programme through their geographic proximity to programme works)
- employees (delivered changes and benefits will directly impact employees of the client organisation)
- shareholders (individuals or legal entities owning shares in the client organisation undergoing change that are affected by the business change)
- end users (those who will ultimately work in new facilities provided or who will be the benefices' of the outcomes of the programme)
- customers (customers of client organisations who will be affected by the business change)
- statutory and regulatory authorities (most programmes will be subject to a range of organisations that will impose restrictions on the way they can be implemented)
- interest groups (the members of which share common interests and control some area of activity, e.g. non-profit organisations and voluntary organisations)

Programmes need to identify and map the stakeholder landscape to engage and communicate effectively. To assess the level of engagement required and the impact stakeholders will have in meeting the objectives of the programme, a stakeholder map (see Figure 1.9) should be developed. This map will define the different tiers of stakeholders according to their potential to affect the reputation or the progress of the project or organisation, with the programme at the centre.

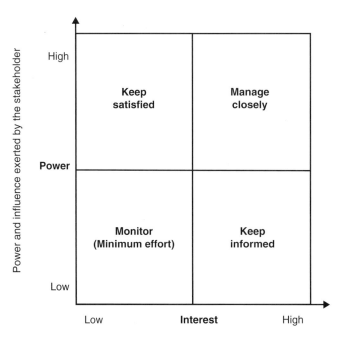

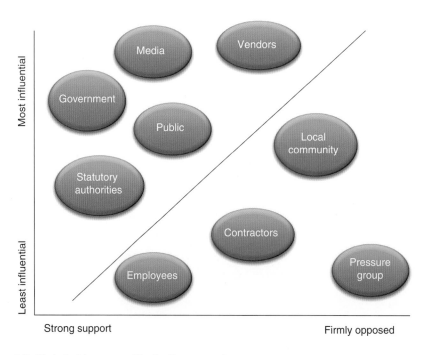

Figure 1.9 Stakeholder map – illustrative example.

1.9 Portfolio management

In the eventuality that there are a number of projects and programmes running in parallel within an organisation, the organisation will often utilise a portfolio approach to govern and administer the initiatives, projects and programmes to identify and manage priorities.

The portfolio management approach will aim to understand the current strategic intent of the organisation and will determine the optimum spectrum of programmes and projects that would provide the most effective and efficient way of achieving the strategic vision by balancing the resources, risk and benefits sought.

Typically, in any organisation portfolio management is an ongoing activity, for unlike most programmes and all projects, it will not have a defined end date.

The administration, management and governance of portfolios follow principles similar to a programme; however, unlike programmes, the procedures are open ended and subject to continual reviews at the highest level of the organisation.

Large organisations may have multiple portfolios, in which case an additional layer of management and governance will be necessary between the senior decision-makers and the portfolio management levels for reporting and administration purposes.

Figure 1.10 illustrates an example of a portfolio management structure in the context of project and programme management.

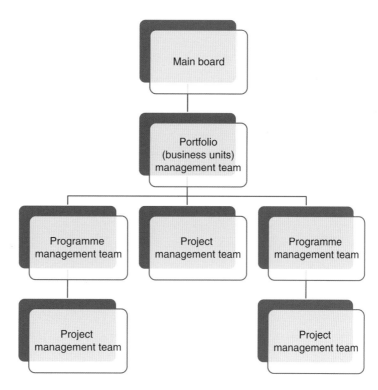

Figure 1.10 Portfolio management structure.

2 Stage A: Inception

> - Are the vision and the change sought set out clearly?
> - Are the benefits expected as a result of the change initiative(s) defined and clear?
> - Have the key resources been identified and appointed?
> - Has initial budget been allocated to proceed with the initial activities?

2.1 Purpose of stage

The purpose of Stage A is to determine whether the strategic aspirations (intent and benefits being sought) of an organisation can be achieved by executing a programme of works.

2.2 Stage outline

During inception, the organisation or organisations interested in implementing the strategic change or undertaking, will articulate the state they desire to achieve in a 'vision statement' (describing what they want to come about). This statement will be used to define what physical outcomes will need to be delivered by a programme in order for their vision to be realised; this document is known as the 'programme mandate' (the plan for the next few years). See Figure 2.1.

Vision statement

The process determining the need for a programme is likely to be complex and lengthy. This is because programmes by their nature are likely to be large enterprises, involving a wide mix of interested parties and large expenditure of capital, resources and effort and creating a large environmental and societal impact.

The inception stage will be undertaken by the sponsoring organisation which itself may comprise a number of separate legal entities.

At a senior executive level in the sponsoring organisation, there will be consideration of the need for a business change, or for the creation of a new enterprise or capability, that arises out of the strategic business objectives of the sponsoring organisation.

Code of Practice for Programme Management in the Built Environment, First Edition. The Chartered Institute of Building.
© 2016 John Wiley & Sons, Ltd. Published 2016 by John Wiley & Sons, Ltd.

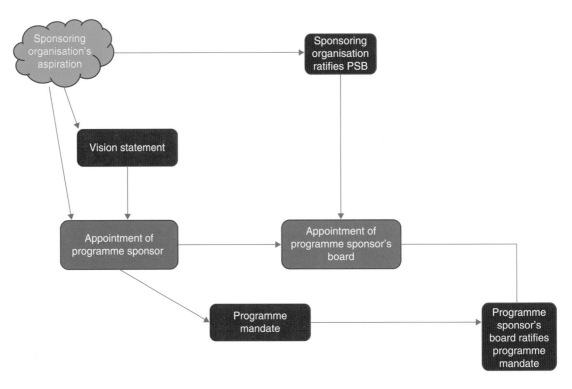

Figure 2.1 Stage A: Inception.

This corporate aspiration is described by a vision statement, which is produced at an executive level and sets out the intent and benefits being sought. It is likely to be subject to a board-level approval and authorisation.

A vision statement can also be described as business vision for change as it is at the root of programme design and benefit definition (see Appendix T1 for a vision statement template).

Its key attributes can be summarised as below. They are:

- focused: simple, clear, concise
- motivational and inspirational
- feasible, with realistic and achievable goals
- easily conveyed ('the elevator pitch')
- at a high level, with flexibility to be developed further for each audience
- unambiguous and collectively understood by all in the same way
- free from technical jargon and uses common and plain language

Examples

- Major transport programme: To build a world-class new railway which expands the capital's infrastructure
- Major sporting event: To host an inspirational, safe and inclusive Olympic and Paralympic Games and leave a sustainable legacy for London and the UK
- Major energy generation programme: Leading the energy change by developing and implementing a carbon capture and storage (CCS) programme in UK

Stage A: Inception

Programme mandate

One of the first actions of the programme sponsor is to expand on the vision statement, setting out in greater detail what the programme needs to achieve in terms of outcome(s) and what the programme seeks to deliver. Depending on the nature of the programme, defining this may be obvious and straightforward or it may be highly complex, requiring development in association with parts of the sponsoring organisation and possibly with external parties. This process describes the programme mandate (see Appendix T2 for a template for a programme mandate).

2.3 Stage organisation structure

2.3.1 Stage structure and relationships

The inception stage requires the appointment of the programme sponsor by the sponsoring client body; the programme sponsor has total overall responsibility for the delivery of the programme and for the achievement of the required outcome. Adding detail and clarity to the initial aspiration is carried out principally by the programme sponsor. During the course of this stage, the programme sponsor will have acquired sufficient knowledge of the requirements of the proposed programme to determine the appropriate membership of the programme sponsor's board (PrgSB). The PrgSB will provide strategic direction to the programme and will be the body ratifying key approvals and decisions on behalf of the programme business partners. (See Figure 2.2 for a chart of the Stage A organisation structure.)

2.3.2 Stage roles of key participants

Client/sponsoring organisation

The client body that has decided to embark upon a major business change selects and appoints a suitably experienced individual who has the business, technical

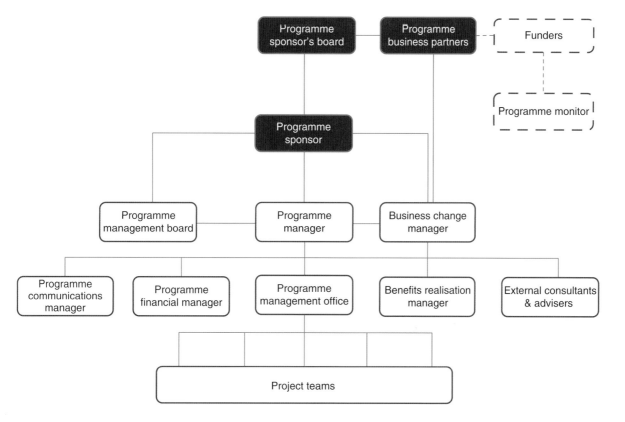

Figure 2.2 Stage A: Inception – Organisation structure.

and managerial knowledge and skills to direct the successful delivery of the anticipated outcomes. Appointment of the programme sponsor should be accompanied by the client's production of a vision statement setting out what the programme needs to achieve.

Business partners and funders

Those organisations that have a business or financial interest in the outcome of the programme assist the client in developing the vision statement and in providing advice to the programme sponsor in developing the programme mandate.

Programme sponsor

Announcement of the vision statement should be accompanied by the appointment of the programme sponsor. Having determined the need for a programme, it is essential for the sponsoring organisation to rapidly nominate one person from within the organisation to take ownership of the programme. The programme sponsor has full accountability for directing and leading the programme and for its delivery.

For very large programmes, it may necessary for the role of programme sponsor to be provided by a group or even by an organisation. In this situation there still must be a lead role that carries the ultimate responsibility.

The focus of the programme sponsor's activity during inception is interpreting the outcome set out in the vision statement regarding what actually needs to be accomplished by the programme in order to achieve this outcome; this is defined in the programme mandate.

Key activities addressed by the programme sponsor to produce the programme mandate include the following:

- Reviewing the vision statement with the client
- Reviewing the vision statement with the programme business partners
- Breaking down the overall outcome into achievable objectives
- Engaging with external parties/key stakeholders who have specialist knowledge that helps to inform the programme mandate
- Determining the way the programme will be managed
- Preparing the programme mandate document

Developing the programme mandate should allow the programme sponsor to determine the functions and/or individuals to comprise the PrgSB. In relation to the PrgSB, the programme sponsor needs to:

- make recommendations to the sponsoring client body on the proposed composition of the PrgSB
- develop terms of reference (ToR) for the function and operation of the PrgSB
- oversee the formation of the PrgSB
- hold an induction meeting with the newly formed PrgSB

Following the production of the programme mandate, the programme sponsor has further activities to undertake:

- Presenting the programme mandate to the PrgSB
- Developing a ToR and plan for carrying out Stage B: Initiation

Stage A: Inception

- Securing the approval of the PrgSB for the programme mandate
- Securing the approval of the PrgSB to proceed to Stage B

Programme sponsor's board

Once formed during the inception stage, the PrgSB will be asked by the programme sponsor to review and ratify the programme mandate, together with the terms of references and time schedule indicating how the next stage will proceed. PrgSB approval allows the programme sponsor to proceed to the next stage of the programme. During inception, the PrgSB undertake a number of activities:

- Ratifying the selection and appointment of the business change manager
- Providing advice and input into the programme mandate
- Resolving issues raised by the programme mandate
- Ensuring the proposals contained in the programme mandate are consistent with the requirements of the functions/organisation that they are representing
- Reviewing and give approval to the programme mandate
- Giving approval to proceed to Stage B

2.4 Programme management practices

2.4.1 Strategic change

Change management is about doing things better: delivering organisational changes while managing quality and risk, as well as producing measurable and sustainable benefits adding value to operating business models. Programme management is a systemised and structured approach designed to realise strategic objectives and manage the risks of delivering programmes.

Programme management brings reality and method to the aspirations of strategic planning and to the delivery of strategic change with the right strategic objectives phased and delivered 'on cost, on time, on quality and with benefits'.

A programme is a temporary organisation designed to operate, learn and adapt in a complex environment of interrelated projects, people and organisations. In this context, the programme manager is the chief executive officer of a temporary organisation with 'the ability to carry the flame for what users want'.

Programme delivery in context of the built environment is illustrated in Figure 2.3. It is a circular process running from strategy to benefits realisation, framed between (i) businesses as usual – strategizing and realising benefits while operating a business/an organisation – and (ii) programme environment – planning and managing a programme to deliver outcomes and benefits.

Strategic change: setting the direction

Strategic planning is typically carried out at the organisational governance level. It is at this level that the strategic direction for the organisation is set, thus strategic planning typically is not a programme management practice. Rather, strategic planning establishes the direction and strategic changes for an organisation's programmes. Normally initiated at a senior level, it is a process which defines the business intent, formulates the vision statement and captures the strategic changes to be delivered via a transformational programme that will have positive economic, environmental and social impacts.

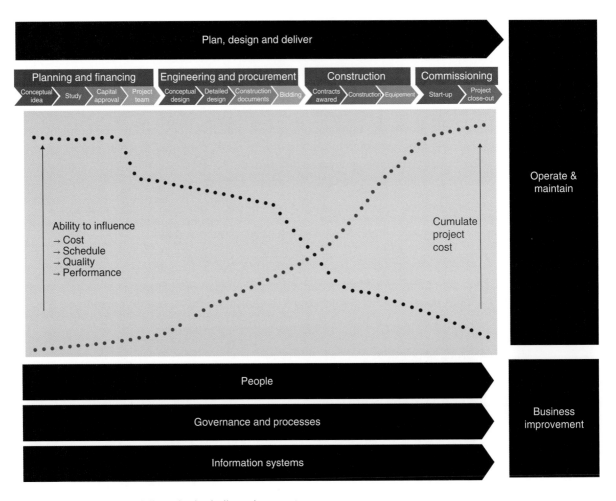

Figure 2.3 Programme delivery in the built environment.

Strategic changes are defined and selected through the strategic planning process and the analysis of macro-environmental scenarios in order to produce qualitative and quantitative benefits, adding value by the expected return of this investment. For each environmental scenario driving a strategic change, the following questions will be considered:

- What political, environmental, social and technological trends are you noticing?
- What are the related issues or challenges?
- What advantages or opportunities are there?
- What impact might these have on the organisation?

A number of examples are listed below for each type of change:

- To enter new geographical market
- To renew equipment fleet
- To become leader in use of building information modelling (BIM) technology
- To develop strategic capability in programme management
- To implement operational waste regulation
- To implement carbon emissions regulation
- To adopt new licensing requirements

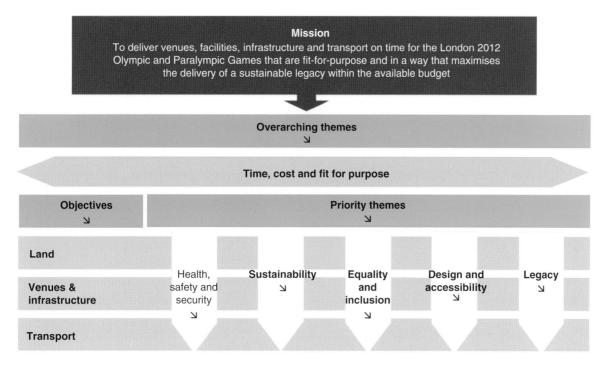

Figure 2.4 Olympic Delivery Authority – London 2012. (http://learninglegacy.independent.gov.uk/)

- To increase passenger capacity
- To deliver a major sporting event
- To improve flood controls
- To deliver alternative energy generation sources
- To reduce patient waiting time for hospital admission

An overview of the overall strategic approach (see Figure 2.4) adopted by the Olympic Delivery Authority to deliver the 2012 London Olympic Games' physical assets is summarised below. Each project – land, venues and infrastructure, transport – was aligned to overarching and priority themes and a set of objectives established for programme delivery and decision-making purposes against specific reporting metrics. The Olympic legacy programme was designed to be delivered through three separated organisations funded in majority by the UK government as follows:

- Olympic Delivery Authority (ODA) – to design and build the venues for games and legacy purposes
- London Organising Committee of the Olympic and Paralympic Games (LOCOG) – to run the Olympic Games
- Olympic Park Legacy Company (OPLC) – to transition the assets in legacy mode

Business objectives: defining the destination

To deliver a successful programme, managers need to understand and consistently communicate their organisation values and vision for strategic change. They also need to translate these into business objectives that are SMART (Specific, Measurable, Achievable, Realistic and Time bound) in order to bring continuous improvement and/or deliver sustainable benefits. The following is an example.

> A building manager has a strategic objective to generate revenue level from a number of vacant sites. At the operational level, construction projects may deliver commercial and residential buildings to the building owner; these can form a programme of works containing multiple projects. Once the projects are completed, the owner has the capability to generate profit which would enable achievement of the programme benefits. Only when the buildings are leased or sold can the benefit be realised and measured. When the expected revenue (as identified during the strategic decision-making stage) has been achieved, the strategic objective is deemed to have been met, thus effecting the completion of the programme.

See Figure 2.5 below for some further examples.

Type of change	Strategic change	Strategic objectives
Internal changes	To enter new geographical market	Enter x new market and generate £x million in revenues by year xxxx
	To renew equipment fleet	Review supply chain and renew xx% by year xxxx
	To become leader in use of a new technology	Use BIM for all projects over £xx m by year xxxx
	To develop strategic capability in programme management	Select and up-skill xxxx senior managers/year
External change	To implement operational waste regulation	Recycle 100% of project waste by year xxxx
	To implement carbon emissions regulation	Reduce carbon emissions by 20% year on year
	To adopt new licensing requirements	Implement new licenses requirements on all sites within x months
End user/client lead (i.e. client led strategic change requiring new physical assets)	To increase passenger capacity	Build new airport, train line, port, etc. by year xxxx
	To deliver a major sporting event	Deliver on time and on budget within scope and safely
	To improve flood controls	Reduce flood risk in x area by xx%
	To deliver alternative energy generation sources	Generate xx% of country energy needs from alternative sources by year xxxx
	To reduce patient waiting time for hospital admission	Reduce waiting time to xx weeks for specific clinical cases by year xxxx

Figure 2.5 Strategic change and strategic objectives by change type.

An effective way to capture strategic objectives and filter down this information in a programme and an organisation's functional areas is illustrated in Figure 2.6. The integration of programme reporting and business reporting metrics underpinned by earned value management is key to a successful delivery.

Business change process: delivering continuous improvement

For customer or client-led programmes (i.e. delivering multiple physical assets or complex long-term contracts), the focus of the following sections, the traditional linear delivery cycle for major projects offers a robust model for delivery. Using this model, an effective programme will anticipate and integrate organisational changes within financial, contractual and physical constraints as time passes. Innovation through clarity of strategic intent and creative design coupled with adapted controls (risk management and change management) is critical to successful scope delivery (on time, to budget and quality) and benefits realisation. The programme and organisational structure (including resource levels) will be gate controlled and will evolve through each delivery phase.

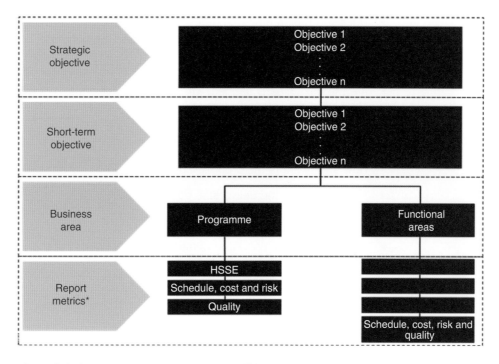

Figure 2.6 Strategic objectives alignment. HSSE – Health, Safety, Security & Environment.

2.4.2 Funding policy and strategy/arrangements

Sources of funding for a programme may come from single or multiple (internal or external) organisations or business units contributing to the programme budget and anticipating a benefit from a strategic change. Not all the projects within the programme may have the same funders, given the potential time variances. Hence, the number and range of funders may change over time; typically, these may include banks, pension funds and insurance companies, together with investors from the UK, EU and other international sources. This is likely to depend on where the individual projects are located. The nature of the projects themselves may attract different types of investors and funding organisations.

The funding policy for the programme typically will depend on the nature of the programme and whether it is in public sector or private sector. At the inception stage, the key decision for the type and nature of funding may not be fully detailed: perhaps, neither would the budgetary requirements or cash flow. Depending on the outputs and outcomes, it is possible that different streams of funding may have to be pulled together for different projects and outcomes.

Some of the initial considerations in relation to funding arrangements may include the following:

- Is the proposed programme a strategic fit for the organisation?
- Does the work require long-term and extensive support?
- Are there adequate considerations for lowering the financial risk, thereby reducing the cost of money?
- Is the proposed programme in an area of high strategic priority for a large investment?
- Is there a case for a major investment in the context of the overall portfolio and budget?

- Does the programme mandate generate confidence that the programme has the potential to successfully manage and deliver the benefits?
- What is the quality of the proposed programme?
- Are the risks clearly defined, with mitigation measures proposed?
- Are there contingency options in place?
- Are the delegated authorities clearly defined?
- How significant is the proposed programme in terms of its potential impact?
- Are there clear and deliverable benefits?
- Is it important to pursue this programme now?
- Is the vision and mandate realistic in its time frames and proposed resources?
- How convincing and coherent is the overall proposed approach?
- Has the organisation demonstrated a clear commitment to the proposed programme and the desired benefits?
- Are there any internal or external dependencies or funding?
- Does the programme represent good value for money?
- Are the proposed arrangements for the stakeholder understanding of this programme appropriate and sufficient?

Not all of the above considerations may be fully determined at this stage; however, these are some of the key criteria (and by no means this is an exhaustive list) that the programme brief (see Stage B) must be able to satisfy in order to secure funding.

3 Stage B: Initiation

> - Does the programme brief develop in detail what is expected in the vision statement and programme mandate?
> - Does the business case set out the feasibility parameters of the deliverables and benefits?
> - Has the funding been identified and secured for the next stage?
> - Have the appropriate resources been identified and appointed?

3.1 Purpose of stage

The purpose of Stage B is to develop detailed proposals for the programme business case to determine what the programme will be able to deliver and make an informed judgement regarding its financial viability.

3.2 Stage outline

The required outcomes established during inception are developed by a more detailed analysis, the programme brief, to identify what would be required from a programme and how it would need to be delivered to secure the outcomes. A programme business case is then compiled to determine the financial viability of the proposed programme. See Figure 3.1.

Identification process

The identification process includes the development of the programme mandate to provide further details of what the programme needs to achieve.

Called the 'programme brief' (see Appendix T3 for a template), this document covers aspects such as the following:

- Statement of intended outcome(s)
- Statement of the benefits required from the programme
- Awareness of how these benefits will be measured
- Consideration of the strategy for delivering the programme
- Indication of the organisational structure necessary to deliver the programme

---

Code of Practice for Programme Management in the Built Environment, First Edition. The Chartered Institute of Building.
© 2016 John Wiley & Sons, Ltd. Published 2016 by John Wiley & Sons, Ltd.

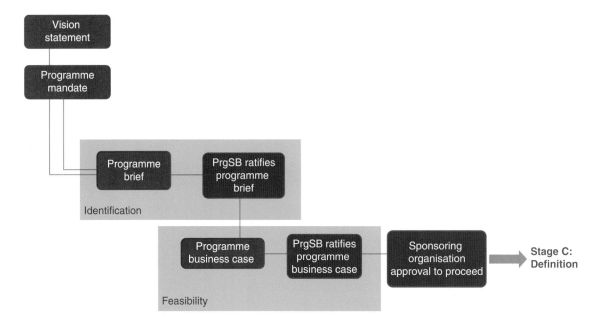

Figure 3.1 Stage B: Initiation.

- Statement of all costs of delivering the programme
- Statement regarding the funding mechanism
- Outline of time schedule for the achievement of the programme, including any critical dates, interdependencies between projects and any external dependencies or dates
- Identification of any external processes, procedures or approvals that will be relevant to the programme
- Statement of any key risks, issues, assumptions and constraints that have the potential to impact the delivery or outcome of the programme and its projects

Who will assist with the preparation of the programme brief, other than the programme sponsor, will vary from programme to programme and will depend on factors such as the sensitivity (commercial, corporate or political) of the programme, the complexity of the programme and the degree of knowledge and expertise available within the sponsoring organisation. In some circumstances it may be possible for it to be produced by the programme sponsor, while in others it may require the assistance of external advisers or the early involvement of some of the key members of the programme management team, such as the programme manager, programme financial manager or head of the programme management office (PMO).

At some point during the initiation stage, it will be necessary for the programme sponsor to have the specialist knowledge provided by somebody who has a thorough appreciation of what the final outcome of the programme needs to be. A business change manager (BCM) views the development of the programme from the perspective of the final end state. The BCM will be appointed from within the sponsoring organisation or by the organisation that will be managing the enterprise being facilitated by the programme.

Programme sponsor's board (PrgSB) approval is required for the description of what the programme is to achieve and what is required to deliver it as set out in the programme brief. This approval allows work to be undertaken after consideration of the feasibility of achieving the programme outcomes.

Feasibility process

The feasibility process involves making an informed study of the effort and costs of delivering the programme's objectives against the returns to be obtained and establishes the financial viability of undertaking the programme. Based on the information contained in the programme brief, an investment appraisal is carried out balancing the expected benefits with the potential risks and threats of delivering the programme. This information represents the programme business case (see Appendix T4 for a template). This is a document that will be used throughout the programme as a control to verify that deliverables being achieved are aligned with the programme objectives.

Preparation of the business case is the responsibility of the programme sponsor, but it is expected that the sponsor will require assistance from financial and investment specialists who have an appreciation of the undertaking.

When the PrgS considers that the business case presents a viable programme, it is submitted to the PrgSB for their review and approval. By signing off on the business case, the PrgSB is confirming they are satisfied the programme can proceed to the definition stage (Stage C). As this approval commits a significant level of resources and expenditure in some instances, it may therefore be necessary to refer the business case to the sponsoring organisation's executive board in order to obtain the instruction to proceed to the next stage.

The PrgSB should also be asked to ratify the terms of reference and time schedule indicating how the next stage will proceed.

3.3 Stage organisation structure

3.3.1 Stage structure and relationships

The key participants in this stage are the PrgS and BCM, who work together to develop the details of the proposed programme such that its viability, in the form of a valid business case, can be demonstrated to the PrgSB and the programme business partners. It is likely that during this process the programme sponsor will require the assistance of the programme manager to ensure that the assumptions made regarding programme implementation are appropriate and achievable. See Figure 3.2.

3.3.2 Stage roles of key participants

Programme sponsor

This stage commences with developing the programme mandate into a more comprehensive programme brief. This process involves the PrgS in the following ways:

- Selecting and appointing the BCM
- Selecting and appointing the PrgM
- Ensuring, with the BCM, that the benefits identified as being delivered by the programme are compatible with the business requirements of the client
- Providing initial consideration of the governance policies for the programme
- Developing the outline strategy for implementing the programme
- Providing initial consideration of likely timescale
- Providing initial consideration of likely cost

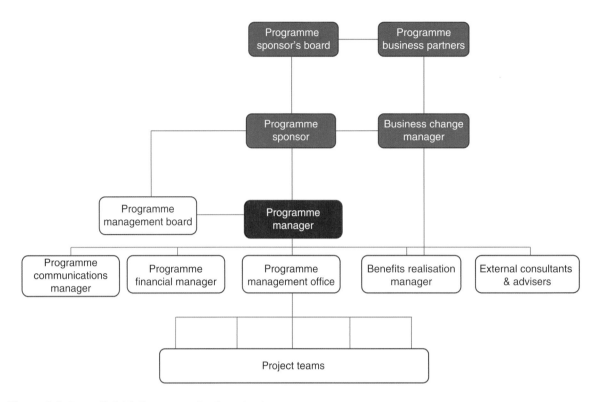

Figure 3.2 Stage B: Initiation – organisation structure.

- Providing initial consideration of major constraints and risks
- Reviewing with the client funding options
- Producing the programme brief
- Presenting the programme brief to the PrgSB
- Securing PrgSB approval of the programme brief

Approval of the programme brief allows the PrgS to proceed with a series of activities related to developing the programme business case:

- Review in conjunction with BCM the identification of the benefits to be delivered by the programme
- Review in conjunction with the PrgM the methodology for delivering the programme
- Review with specialist adviser the available funding options
- Develop a funding strategy
- Review any relevant lessons learned from previous projects or programmes
- Oversee the production of the programme business case
- Present the programme business case to the PrgSB
- Secure approval of the PrgSB to the programme business case

Prior to completion of this stage, and in anticipation of obtaining PrgSB approval, the programme sponsor needs to develop proposals for executing Stage C:

- Develop the terms of reference ToR for Stage C
- Develop in conjunction with the PrgM a time schedule for Stage C

- Develop in conjunction with the PrgM a resources plan for Stage C
- Secure PrgSB approval to proceed to Stage C

Programme sponsor's board

Throughout this stage the PrgSB continue their function of advising, reviewing and approving, which includes:

- providing advice and input into the programme brief
- resolving issues raised by the programme brief
- ensuring the proposals contained in the programme brief are consistent with the requirements of the functions/organisation(s) that they are representing
- reviewing and approving the programme brief
- providing advice and input into the programme business case
- resolving issues raised by the programme business case
- ensuring the proposals contained in the programme business case are consistent with the requirements of the functions/organisation(s) that they are representing
- reviewing and approving the programme business case
- giving approval to proceed to Stage C

Business change manager

During Stage B, the BCM has responsibility for the following:

- Supporting the programme sponsor (PrgS) in the production of the programme brief
- Verifying that the information contained in the programme brief reflects the business objectives of the sponsoring client body
- Defining, based on the objectives set out in the programme mandate, the characteristics and nature of the benefits to be delivered
- Supporting the PrgS in the production of the programme business case
- Ensuring the benefits to be delivered by the programme are clearly stated in the programme business case

Programme manager

The PrgM is introduced into the programme for the first time during this stage to ensure that the information regarding the implementation of the programme is realistic and appropriate. This is a senior appointment and will require a person with high levels of leadership and a proven ability in the successful delivery of programmes and projects. The PrgM's tasks include the following:

- Supporting the PrgS in the production of the programme brief
- Supporting the PrgS in the production of the programme business case
- Developing a strategy for the implementation of the programme
- Establishing the deliverables required to achieve the programme's benefits
- Considering an initial listing of likely projects required to achieve the identified deliverables

- Developing a time schedule for programme delivery
- Developing a cost plan for programme delivery
- Developing a risk register for programme delivery
- Developing a resource plan for programme delivery
- Developing in conjunction with the PrgS a ToR, time schedule and resource plan for Stage C

3.4 Programme management practices

3.4.1 Benefits management

A generic approach to programme benefit management consists of three phases (see Figure 3.3):

1. Capabilities phase: building and delivering the programme
2. Transition phase: transferring and operating the asset
3. Benefits phase: realising the benefits

Benefits delivery means achieving the desired outcomes identified in the business case on time and on budget. Its key stages are benefits identification, benefits management and benefits realisation. To deliver benefits successfully, benefits need to be measurable outcomes and fall into one of the five categories as identified in Figure 3.4.

1 – Benefits identification "Why are we doing this?"	2 – Benefits management "Are we on track?"	3 – Benefits realisation "Are we there?"
Defining realistic and measurable benefits, setting ownership, accountability for delivery, and designing measurement indicators.	Sequencing benefits and maintaining focus on timing for delivery. Tracking progress to support decisions to stay on course.	Planning and transitioning for business ownership and integration after closure. Post-implementation review to support ongoing business and continuous improvement.

Figure 3.3 Benefit delivery in three stages.

Category	Benefit Type (some)	Possible measures
Financial	Revenue enhancement Capital expenditure	Increase revenue by 20% Reduce capex by 20%
Operational	Efficiency Effectiveness Quality Innovation	Increase profitability by 10%
Customer	Service Reputation Brand	Achieve 90% customer satisfaction Become market leader for programme management
People	Morale Capability	Reduce staff attrition rate by 20%
External stakeholder	Regulatory	Reduce CO_2 emission by xxx.

Figure 3.4 Benefits categories.

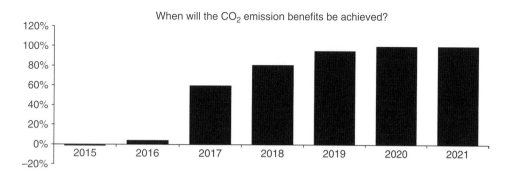

Figure 3.5 Example of graphical representation of benefits realisation over time.

- Stage 1 – Benefit identification methods: 'Horses for courses'. The theme for improvement is a top-down method of articulating the focus areas in which the organisation wants to make a change. They give a clear sense of the areas of change and a broad understanding of where benefits will be seen

  A more detailed approach to identify benefits and their measures is through a tree structure. A tree starts from the driver for change, and through a deductive, analytical and structured approach, processes through to a series of options of potential benefits by category

- Stage 2 and 3 – Benefits management and realisation. Once all benefits captured and defined, they can be held in a central database and summarised to include definition by category and delivery measure and timescale for delivery and owner. For benefit tracking, a graphical bar chart can be used to represent and track benefit realisation over time (see Figure 3.5)

3.4.2 Feasibility study

A feasibility study is an analysis of the viability of an idea. The programme brief will contain the strategic objectives and an overview of constraints and risks, along with high-level financial forecasts; however, to translate the information contained within the brief to a business case, a feasibility exercise will should be undertaken.

The feasibility study should thoroughly examine all the issues and include the following:

- Appraisal of business alternatives with respect to meeting the identified need
- Appraisal of opportunities generated
- Assessment of high-level risks and potential mitigating factors
- Probability of success of delivery, at both programme and project level
- Funding options and scenarios
- Benefits appraisal
- Resources appraisal
- Any fundamental assumptions made and appraisal of the assumptions
- Demonstration of due diligence

It is important to note that once the programme brief is set, in particular for large and complex programmes, a scoping or options appraisal or even a pre-feasibility study may be necessary to narrow down the possible options and scenarios prior to commissioning a detailed feasibility study.

The feasibility study report should consider the programme brief thoroughly and may specifically comment on the following:

- Technical viability
- Financial viability
- Benefits viability
- Contingency options
- Fitness for purpose

The feasibility study report, along with its recommendations, will constitute the backbone of the business case.

3.4.3 Funding arrangements

The initial investigations regarding the nature of funding will essentially determine the potential sources, availability and governance. An initial scoping study must be undertaken to determine whether the availability of funding is in keeping with the forecasted cash flow for the programme.

For private sector projects, commitments must be sought from the fund holders (and the shareholders or directors of businesses) to ensure that there is sufficient funding for the life of the programme. It is often the case that an initial funding is released to kick start the programme, and any further subsequent funding is conditional to key trigger events or achievement of early benefits (typically linked with revenue generation).

In certain instances, it is also advisable to ascertain the provenance of the funding to ensure that it is complying with regulatory and legal requirements.

Large public sector programmes often rely on PPP or PFI (public–private partnership or private finance initiatives) as sources of programme funding. In this scenario, the key steps are the following:

- Implementation of the PPP/PFI/commercial policy
- Managing PPP and PFI projects within programmes
- Controlling the quality of PPP and PFI projects though procurement
- Supporting the transition through operational PPP and PFI projects to ensure that they achieve their benefits
- Managing the market of the operators and investors
- Embedding continuous improvement in PPP and PFI projects and programmes

In order to deliver the above functions, organisations are encouraged to ensure that the following enablers are considered at the initiation stage:

- A clear mandate of roles and responsibilities: There should be clarity regarding who is doing what, what are the delegated limits and who is authorised to make decisions
- Adequacy of resources – capacity and competency: The delivery team has the resources collectively experienced in PFI/PPP, commercial, financial, technical and sector-specific programmes and projects
- Strong relationship – intra- and inter-organisational: Strong relationships are needed within the delivery team and within the key organisations involved

A further option available to public sector organisations is self-financing, where funding could be available through borrowing (provided such borrowing is affordable, prudent and sustainable over the medium term), capital receipts (by selling fixed assets), capital grant (from various central and local government funding, lottery funding, European grants), revenue contributions (albeit under the current regulations the scope is limited) or capital reserves (with funding earmarked for specific capital programmes).

4 Stage C: Definition

> - Have all the activities, deliverables, resources, including enabling technologies and costs, needed to prepare a detailed plan and design for all aspects of the programme been identified?
> - Have all the roles been identified and assigned?
> - Has the governance plan been prepared and put in action?
> - Does the programme sponsor board accept the activities and the costs required to complete the design phase?
> - Is there enough information to identify the outputs, outcomes and deliverables?
> - Does the programme delivery plan give confidence that the capability will support the realisation of the vision and the benefits?
> - Has the extent and nature of the change required to achieve the vision been identified?
> - Is there confidence that the actions that have been put in place to resolve any outstanding risks or issues and support a decision to continue to proceed to the next stage?
> - Does the programme sponsor board accept the programme design?

4.1 Purpose of stage

The purpose of this stage is to provide a detailed definition of how the programme is to be set up, the delivery strategy, the level of resources that delivery will require and the governance controls to be applied, collectively known as the programme delivery plan (PDP).

4.2 Stage outline

Approval of the programme business case authorises the appointment of the principal members of the programme management team, who will then develop the strategy and methodology for the execution of the programme, collectively documented in the PDP (see Figure 4.1).

During the Definition Stage, the elements discussed next must be addressed.

Code of Practice for Programme Management in the Built Environment, First Edition. The Chartered Institute of Building.
© 2016 John Wiley & Sons, Ltd. Published 2016 by John Wiley & Sons, Ltd.

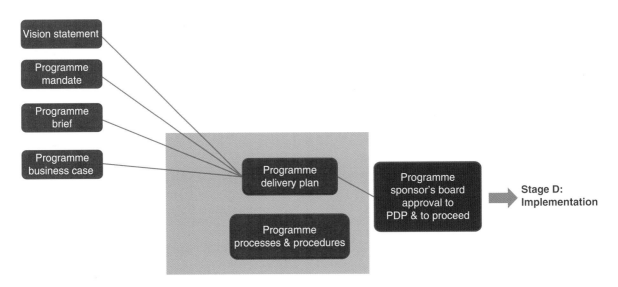

Figure 4.1 Stage C: Definition.

Benefits profiles

The vision statement and programme brief are reviewed in order to develop a greater understanding of each benefit. Benefit profiles are developed, which consider the following:

- A description of the benefit
- The way it will be realised
- How and when this will be achieved or measured
- What measurement mechanism is appropriate
- Dependency on, or relationship with, any other benefit or activity.

The benefit profiles are collated into a benefits realisation plan, which describes how and when benefits resulting from the programme will be obtained

Scope definition and projects register

Preparation of the benefit profiles will enable the development of a project list known as the 'projects register', which will be necessary to achieve the programme outcomes. Each identified project needs to be fully scoped, together with the criteria of output, cost and time established. Any across-project dependencies or external constraints need to be highlighted. Delivery strategies for the projects must be stated, along with the managerial resources required and an indication of where these resources will be sourced.

Scope of undertaking

Scope definition is among the key critical success factors often cited by senior managers. These include:

- Scope clarity and innovative design to create value and save time in delivery
- Culture and diversity exemplified by a core team with no-blame culture from the start and with different backgrounds to provide different perspectives
- Strong risk management processes and accountability (intelligent client model to drive value from supply chain)

Scope definition is fundamental to an effective programme control function in order to build the right organisation and set of policies, procedures and processes for delivery. It will guide the programme to:

- Establish a baseline for future reference and realistic standards
- Monitor performance
- Keep the plan under constant review and take action when necessary to correct the situation

$$Scope \approx f(time, cost, quality)$$

It provides a comprehensive summary of a programme, detailing the scope, schedule, budget and risk based on a clear map of the work and how it is broken down in time and space using the WBS.[1]

A collectively developed, shared, understood and used WBS or activity schedule is at the heart of programme success.

Also needed is a network of activity that clearly sets out the overall strategy for breaking down the programme that can also be supported by a visual schematic geographical representation to show:

- Discrete stages of the programme and projects including any overlap of stages
- Sub-projects, work streams such as information technology (IT) or infrastructure
- Principal or summary activities
- Milestone and key dates
- Inter-relationships between the activities and sub-projects
- The objective to be achieved at each stage

Stakeholder analysis

As programmes tend to be large and are generators of significant change, their influence has an impact on a wider scale than just one-off projects. Because of their nature, it is also likely that programmes have a higher visibility than do projects. It is therefore essential to the design stage that there is a detailed deliberation of the identification, interest and influence of stakeholders; it is likely the list of stakeholders will be substantial. The large impact of some programmes means there is likely to be higher degrees of opposition or dissatisfaction. As effective stakeholder analysis and management is critical to programmes, a full time stakeholder manager, or even a team, may be assigned to ensure this important aspect is dealt with successfully.

A detailed stakeholder analysis and a stakeholder management strategy will be developed from the initial considerations included in the programme business case. A 'programme communication plan' will be created that defines how each stakeholder will be dealt with and the nature, extent and frequency of communications within and outside of the programme team and stakeholders.

---

[1] WBS stands for work breakdown structure, a hierarchical and incremental subdivision of elements necessary to achieve the end objective.

Risks, issues, assumptions and constraints

The programme mandate and brief will have identified a number of risks, issues, assumptions and constraints, and these are now revisited in light of further understanding of the programme that is emerging. This is an opportunity to take advantage of a wider consultation on these aspects.

Risks (circumstances that may occur or change, some identified and expected, others unknown and unexpected) will be analysed in a risk register, and a methodology will be adopted for their ongoing monitoring and management. Some financial contingency allowance needs to be calculated for the impact of the risks.

Similarly, issues (things that have already happened and need action) will be listed in an issue register and a methodology adopted for their ongoing monitoring and management.

Initial assumptions need to be reviewed to clarify likely impacts on the programme. Some assumptions may concern factors outside the direct influence of the programme, and these may need to be referred to the project sponsor who may in turn need to obtain further advice from a higher level in the sponsoring organisation. Any additional assumptions made during this stage also should be considered and included on the register. Where appropriate action plans are developed, a methodology for regularly reviewing assumptions also should be developed.

Any constraints need to be subjected to a similar approach. Consideration should be given to devising strategies that avoid major constraints. Constraints may be created by:

- Operational requirements
- Funding issues
- Commercial and political sensitivity
- Regulatory and legal requirements
- Timing issues to do with availability (legal agreements, ownership, resources)
- Key calendar dates

Programme timescale plan

Based on the information contained in the programme brief; the benefit profiles; projects register and risks, issues, assumption and constraints registers, an overall delivery time schedule for the programme is produced. [Note: In common with other CIOB publications, to avoid confusion between terms, 'schedule' is adopted throughout this publication rather than 'plan' or 'programme'.] The time schedule estimates anticipated completion of the programme, the anticipated duration and timing of individual projects, and identifies when benefits realisation are expected. It highlights dependencies between projects and any external dependencies. In situations where there is a complex pattern of dependencies it may be helpful to produce dependency charts.

It should be possible to identify the critical path sequence through the programme time schedule as this is helpful information when having to make decisions on resource priorities. It should be acknowledged that, at this stage, it may not be possible to identify all the projects that are required to achieve the programme's objectives.

In addition to the overall programme time schedule, a more detailed start-up time schedule should be produced describing the timing of activities necessary to mobilise the next stage of the programme once approval to proceed has been obtained.

Programme financial plan

At the point at which the delivery plan and delivery time schedule have been devised, it is possible to prepare a financial statement that collects all the costs that have been identified in relation to implementing the programme.

The financial statement should be compared with the financial information contained in the programme business case and any significant variances highlighted.

From the financial plan and the delivery time schedule the expenditure cash flow profile can be calculated. This is a critical document because it informs the funding commitment for the programme. It must therefore be presented, discussed and approved by a senior executive of the sponsoring organisation, such as a financial director, before it can be included in the PDP.

Compared with a project budget, there is a higher degree of uncertainty attached to determining programme budgets as the timescales involved are likely to be much longer and the exact number and scope of projects to be instructed may not be known until later in the implementation of the programme.

The high cost of many programmes creates a substantial risk to the financial standing of the sponsoring organisation; therefore, oversight of the financial aspects of programmes is a crucial function. The role, of programme financial manager must be undertaken by an experienced financial manager who has expertise in dealing with complex issues around tax liability, capital allowances and programme funding.

Transition plan

Prepared principally by the business change manager (BCM), a detailed strategy must be produced to describe the mechanism by which each project output is handed over and incorporated into the final enterprise. How this is achieved will vary depending on the nature of the undertaking. In some situations project outputs can be immediately put into a 'business as usual' state, but others may need an assembly of numerous 'outputs' from projects from different programmes.

The 'transition plan' will explain what management infrastructure must be in place from the new enterprise in order to take ownership of project outputs as they become available. The document should also set out the acceptance criteria for each of the deliverables.

Programme delivery plan

Production of the documents discussed above, as identified earlier in Figure 0.2 and Figure 0.3 in the Introduction, collectively form what is referred to as the PDP. (See Figure 4.2.) The PDP is a detailed description of what the programme will deliver, how and when it will be achieved and the financial implications of its delivery.

The PDP is presented to the programme sponsor board (PrgSB) for their consideration and, providing it falls within the parameters established in the business case, their endorsement that the programme can proceed to implementation.

In situations in which the sponsoring organisation either has a complex structure or comprises a number of separate legal entities, or where stakeholders have a critical influence on or are significantly affected by, the programme, the PrgSB may require further consultation on, or consent to, the PDP prior to their being prepared to authorise implementation.

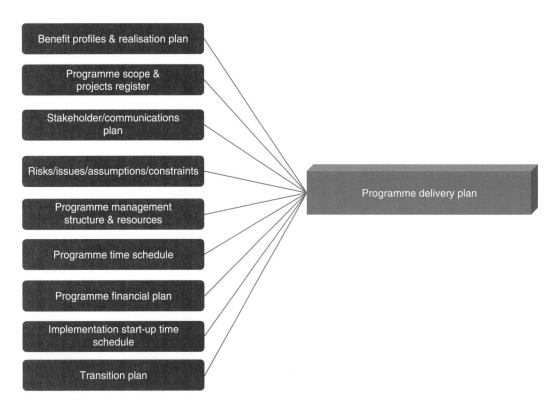

Figure 4.2 Contents of the programme delivery plan.

4.3 Stage organisation structure

4.3.1 Stage overall structure and relationships

Stage C marks the commencement of the involvement of almost the full programme team as they collectively evolve the programme's implementation strategy and fully plan how the programme will be executed. See Figure 4.3 for a structure diagram. Control of the progress of the programme, which up to this point has been with the programme sponsor (PrgS), moves to the PrgM, who now has prime responsibility for planning and designing the way the programme will proceed. Reporting to the PrgS, the PrgM is assisted by a range of personnel: a PrgFM, a programme stakeholder/communications manager and a team of specialists within a programme management office (PMO).

The BCM, working in conjunction with the PrgS and PrgM, continues with the function of ensuring at each stage that what is being proposed or delivered matches the requirements of the client. The business change manager takes responsibility for benefits management and for beginning to develop plans for the transition of the undertaking into its finished state.

In accordance with good management practice, the production of a roles and responsibilities matrix is a helpful device for assisting the determination of the level of structure and resource required.

4.3.2 Stage roles of key participants

Programme sponsor's board

The PrgSB have a number of key roles during this stage:

- Provide advice in relation to the PDP
- Review the PDP

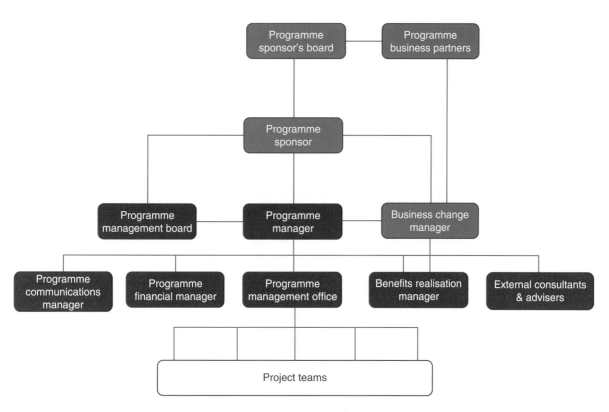

Figure 4.3 Stage C: Definition – organisation structure.

- Resolve any issues raised by the PDP
- Review and give approval to the PDP
- Give approval to proceed to Stage D

Programme sponsor

Having appointed a PrgM to take responsibility for this stage, the PrgS adopts a more strategic role and carries out the following:

- Acts as the interface between the programme management team and the client body
- Provides direction and advice to the PrgM
- Selects and appoints, in conjunction with the PrgM, additional members of the programme management team
- Resolves any queries, ambiguities and conflicts with the PrSB
- Determines the change control process
- Reviews and approve the PDP

Business change manager

During this stage, the BCM continues to focus on the benefits to be delivered and the final outcome of the programme by doing the following:

- Selecting and appointing a BRM (if required)
- Developing benefit profiles to match the outcomes defined in the programme brief
- Developing transition plans

- Confirming acceptance criteria for the programme deliverables
- Reviewing and confirming, in conjunction with the PrgM, the number and nature of projects required to achieve the deliverables

Benefits realisation manager

Depending on the size of the programme, it may be necessary to appoint a benefits realisation manager to support the BCM. The role of the benefits realisation manager during this stage is to:

- Develop the benefit profiles
- Assemble the benefits realisation plan
- Determine the mechanism for realising and measuring benefits
- Determine the critical success criteria
- Determine the acceptance criteria for programme deliverables

Programme manager

Leading this stage, the PrgM has a number of key tasks:

- Identifying the roles of the programme management team that need to be available during this stage
- Selecting and appointing, in conjunction with the PrgS, the remaining members of the programme management team required during this stage [Note: Depending on the individual circumstances these may be internal or external appointments].
- Defining the full scope of the programme
- Arranging for a physical location for the programme team
- Arranging for the establishment of the IT infrastructure to support the programme team
- Overseeing the production of the PDP
- Reviewing and confirming the programme's time schedule
- Reviewing and confirming the programme's cost plan
- Reviewing and confirming the programme's risk analysis and risk register
- Determining, in conjunction with the programme information manager, the information/document system to be adopted
- Reviewing and confirming the programme's governance policies and procedures
- Reviewing and confirming the programme's financial policies and procedures
- Reviewing and confirming the stakeholder analysis and communication plan
- Determining the number and scope of projects necessary to achieve the programme's deliverables

Programme financial manager

The PrgFM has responsibility for ensuring that the programme budget determined during this stage aligns with the business case. The financial manager's principal activities include:

- Overseeing the development of the overall programme budget
- Ensuring the budget conforms with the sums contained in the business case

- Determining expenditure cash flow forecast
- Determining the risk and contingency allowances to be included within the programme budget
- Confirming the programme's funding arrangements with the PrgS and PrgM
- Reviewing the cash flow forecast with the funders
- Defining the programme's financial policies and procedures
- Liaising with the financial director(s) of the sponsoring client on matters relating to the implication of the programme on their financial reporting and tax affairs
- Ensuring prudent financial governance

Programme management board

The programme management board (PrgMB) comprises the senior managers of the programme management structure and provides advice and support to the programme manager. The PrgMB should meet regularly to review the programme's progress and to highlight and resolve any issues that may be hindering it.

The PrgMB is likely to be composed of the following managers:

- Programme sponsor
- Programme manager
- Business change manager
- Stakeholder/communications manager
- Programme financial manager
- Head of the PMO together with project managers from key projects as and when required

Stakeholder/communications manager

The stakeholder/communications manager's role includes establishing an understanding of the programme's stakeholders and a plan determining either their level of engagement or how they will be kept informed of the progress of the programme. This will involve the following:

- Identifying all stakeholders
- Carrying out an influence/impact analysis
- Developing a communication plan for stakeholders
- Developing a communication plan for staff within the client organisation affected by the programme
- Developing a public relations plan for the programme

Programme management office

In addition to requiring a PrgM, this stage requires that the core members of the PMO also be available. These include the head of the PMO and sufficient staff with the necessary technical expertise to carry out the functions required during this stage to develop the PDP and to establish the governance controls. The size and range of expertise required will depend on the nature and complexity of the programme. In some instances the input required from the programme team may be available from within the sponsoring organisation, but in many others it is likely that all the expertise will be externally sourced specifically for the programme. In addition to specialist input on matters such as legal aspects, property and finance, some sponsoring organisations may need to contract with organisations to provide all the programme management delivery capability. These positions include the following:

- Head of PMO: Responsible for setting up the PMO and for ensuring all systems, processes and procedures are established in readiness for the implementation of the programme

- Scheduling manager: Responsible for establishing the planning and schedule infrastructure to be adopted across the programme and its projects and for developing the overall programme schedule

- Cost manager: Responsible for establishing the cost management system and protocols to be adopted across the programme and its projects and for developing the cost budget and cash flow forecasts

- Risk manager: Responsible for establishing the risk management system to be adopted across the programme and for carrying out the process to determine a risk analysis for the programme

- Document/information/IT security manager: Responsible for determining, in conjunction with the PrgM, the document and information system to be rolled out across the programme and for installing this system in preparation for implementation, including ensuring a secure information storage and exchange mechanism

- Data manager: Responsible for data collection, data system management, data reporting and analysis and ensuring data collaboration among all the participants in data exchange

- Administrator(s): Responsible for providing sufficient support to allow the efficient and effective operation of the PMO

- Health and safety manager: Responsible for setting up a system that is appropriate for the health and safety regulations and legalisation existing in the environment(s) in which the programme is being carried out

- Sustainability and environmental manager: Responsible for establishing from the programme brief the sustainability drivers, identifying any statutory and regulatory requirements affecting the programme, and developing a framework for sustainability targets

- Specialist advisors: During this stage it is likely that the project management team will need to obtain the advice and assistance of specialist advisors. These areas could be legal, real estate, financial or a range of technical aspects

Establish programme organisation

- Appoint the remainder of the PMO and any other members of the programme management structure not in place

- Arrange for establishment of a physical location for the programme team

- Establish the IT infrastructure to support the programme

- Set up the programme-wide systems, such as intranet sites and information management (building information modelling [BIM])

Benefits management

- Continually review the contribution project outputs are making towards benefit realisation
- Review benefit realisation against the key success criteria
- Regularly review the benefit profiles in the light of greater knowledge and understanding of the programme, and any changes to the programme's objectives
- Keep aware of any changing external circumstances that may affect the outcomes of the programme and impact benefits

Change management

- Manage the process of appraising changes as instructed by the PrgS
- Advise the PrgS and PrgSB of the implications for the programme and its deliverables of the changes incorporated

Audits

- Depending on the nature and complexity of the programme, it may be necessary for the PrgS and/or programme manager to initiate either internal or external audits of the programme or projects

Transition

- BCM incorporates project outputs into the business operations of the new enterprise in accordance with the transition plan

Stakeholder analysis

Apart from the stakeholders in Chapter 1, Section 1.7.5, a programme will often include other stakeholders that in one way or another may be affected by the programme. Common stakeholder groups include the following:

- General public (people who are only indirectly affected by a programme but who may have a significant influence on its realisation)
- Community (people who are directly affected by a programme through their geographic proximity to programme works)
- Employees (delivered changes and benefits will directly impact employees of the client organisation)
- Shareholders (individuals or legal entities owning shares in the client organisation that is undergoing change who are affected by the business change)
- Customers (client organisations whose customers will be affected by the business change)
- Interest groups (members who share common interests and control some area of activity, such as non-profit organisations and volunteer organisations)

4.3.3 External environment and relationships: mapping the landscape

Any programme affects the environment that it operates in and at the same time is also affected by the environment. The external environment influences business strategy and the success of a programme.

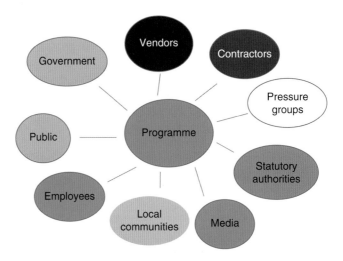

Figure 4.4 Stakeholder map.

It is therefore essential not only to identify external drivers in order to shape a strategic change and a programme, but also to identify, map and track programme stakeholders, as this will help develop support and manage programme communications. See the stakeholder map in Figure 4.4.

Stakeholders include persons and organisations that have an interest in the strategy of the organisation and programme and have an impact on or are impacted by a programme. Stakeholders normally include shareholders, customers, contractors and vendors, staff and the local community.

Programmes and organisations need to be able to identify their stakeholders and judge the level of power they hold to affect the decisions and outcomes of the programme. A first step for this process will be to create a stakeholder map. This map will include all the stakeholders for his or her organisation with the programme at the centre. Stakeholders include:

- Internal stakeholders: Members of the organisations and those with an economic or contractual relationship with the programme
- External stakeholders: Those with interest in the organisation and programme activities or that might be impacted by the activities in some way. Key examples are governments, public, interest and pressure groups, media and news organisations, local communities and statutory authorities

4.4 Programme management practices

During the definition stage, concurrent with the PDP being developed, work is underway on the assembly of a series of processes and procedures that will describe how the various management governance controls will be implemented for both the programme and individual projects. They provide all the programme participants with a consistent understanding of how the PrgS and PrgM, or indeed the sponsoring organisation, want the programme and projects to be managed.

Processes and procedures need to be developed for the areas discussed next.

4.4.1 Scope management

In the context of programme management, scope for a programme relates to the collective objectives of all the component projects and activities. The responsibility for managing individual project scopes will stay with the project managers; however, the PrgM will be responsible for the overall scope management.

Scope for a programme is underpinned by three key components:

- Scoping of information from initiation documentation as prepared in the earlier phases, such as the vision, mandate, brief and business case

- Definition of requirements, which is effectively a detailed design of programme deliverables

- Delivery of requirements, which continues from the programme design and defines the scope of work for each activity or project

The delivery requirements collectively provide a view of the end state of the programme and must align with acceptance criteria to ensure that the project activities address and satisfy the specific requirements.

The programme initiation documentation outlines the current state and the vision for the future state; it is the scope that defines how the programme will enable the changes to ensure that the transition is effected from the current state to the desired future state.

Where the end state is well understood and has a tangible output (e.g. in construction and engineering), it is usual to define the scope as accurately as possible at the beginning. This potentially reduces the level of changes that may be required and keeps costs from escalating. It is also useful to define what is outside of the scope to avoid misunderstandings. Clearly defining what is in and out of scope reduces risk and manages the expectations of all key stakeholders.

Where the objective is less tangible or subject to significant change, for example, business change or some IT systems, a more flexible approach to scope is needed. This requires a careful approach to avoid escalating costs.

It is important to recognise that, particularly for large and complex programmes, it is most likely that initial scoping will be required to undergo changes as risks, issues, and changes in the wider landscape emerge. The programme design must enable a robust and effective change management process to deal with the scope variations and changes, and most importantly, must be capable of identifying and flagging where a change in scope occurs.

It is the responsibility of the PrgM to flag any changes to the scope (be it at project level or programme level) to the PrgS and all changes to scope must be authorised by the PrgSB.

In practice, the majority of scope change requests will be generated at the project level; it is critical that scope change approval is not done at the project level. Those involved in projects can see their own work, but they can't see the interdependencies that exist between projects. Therefore, those working on the projects don't have the right level of understanding as to the impact of scope change requests across the projects. This requires that project scope changes be escalated to the programme level for a decision.

Scope is typically managed both at programme and project level.

- Programme scope: This is owned by the PrgM, and changes to programme scope are managed at the programme level. This may lead to changes necessary at project levels as well

- Project scope: The PDP will identify the processes required at the project level in terms of managing scope. Project scope changes must be examined at the programme level to ensure they are monitored and actively managed. These project scope changes must be elevated to the programme for approval. The impact of an approved scope change request is communicated to the affected projects for management

Managing programme and project scope change is one of the primary responsibilities of the PMO.

These methods are used to identify the activities that are necessary to achieve the programme's objectives:

- Work breakdown structure
- Scope statement
- Acceptance criteria
- Exclusions from scope
- Assumptions

4.4.2 Benefits management

At this phase, it is necessary to define how benefits will be managed, from identification through to realisation, in alignment with the programme vision and the business case.

The key considerations at this phase will include the following:

- How will the benefits be identified – for example, through benefits mapping workshops
- Who needs to participate – typically PrgM, BCM and relevant stakeholders and participants from the programme delivery team
- How will the benefits be monitored/reviewed and when – It is important that benefits are regularly reviewed as the programme delivery progresses to allow for requirement changes, risks, issues and changes to the initial assumptions
- What are the key criteria for reviewing – for example, is there focus on the right benefits, can the targeted value of each benefit be improved, can the costs associated with each benefit be reduced, are there any additional benefits that can be targeted, are the risks and issues associated with benefits being dealt with

The considerations and outputs generated at this phase must be subject to a change control process, and benefit profiles can be used to capture any changes or amendments. The benefit realisation activities in the PDP will also need to be amended to reflect any approved changes.

4.4.3 Risk Management

Risks will be identified, recorded, monitored and managed in the following ways.

- Risk management methodology
- Risk register
- Risk review process

Risks, issues and opportunities (RIO): managing uncertainty

As technical and commercial issues get more complex and financial metrics tend to proliferate, finance and programmes often end up measuring performance and risks in different ways, using various sources of information and, in many cases, using a different language. Risk assessment is not an exact science, and there are a number of different methods to measure or quantify risk.

Effective risk management begins with realism – seeing things as they are – and continues with a joined approach and common understanding at all levels of an organisation or programme. A strong corporate risk management culture and consistent risk-rating methodology are fundamental to the success of a programme in order to focus on the risks that matter.

The guidance below offers a structured common sense-based approach to risk management.

Success factors

The key success factors are as follows:

- Identification and control of risk
- Alignment to business value drivers
- Awareness of changing risk profile and risk appetite
- Comprehensive approach to risk management

Risk categories

Risks will generally fall under the following categories:

- Financial risks
- Reputational risks
- Health and safety risks
- Operational risks

Objectives

The objectives of effective programme risk management are as follows:

- Provide a mechanism for the early identification and resolution of risks that may arise
- Ensure that risks are escalated to and mitigated by appropriate levels of management
- Provide for the visibility of risks that may affect or are affecting high-priority projects
- Provide accountability for the mitigation of project risks
- Provide guidance for the correct control and administration of the recorded risks
- Provide a basis for determining the level of financial contingency required for the programme

Definitions

Risk is often defined as 'an effect of uncertainty on the objectives[2]'. Risk can be categorised as the following:

- Hazard – risk of adverse events
- Uncertain outcome - not meeting expectations
- Opportunity – exploiting the upside

---

[2] Definition obtained from ISO 31000:2009 – ISO guide 73.

An issue may be defined as:

- Any unresolved problem which could jeopardise programme outcomes
- A risk that has materialised
- An uncertainty, which was not previously raised on the risk register, has occurred

It is imperative that all issues are recorded on the project issue log; however, an issue should be reported to the PMO only if the project team cannot resolve the issue or if the project manager determines that the issue impacts other projects that are outside the PMO's scope of responsibility. The issue process implemented by the PMO is to ensure that the appropriate management team is engaged to help resolve problems.

At the portfolio level, risk management and change control is supported centrally, which brings together a set of essential functions to support the successful delivery of programmes and projects, including:

- Coherent upward reporting to the management board on its key programmes and projects to support effective decisions
- Timely sharing of information and lessons learned through outward relationships ... and beyond
- Inward support to help deliver programmes and projects with the right expertise when needed

Quantitative risk assessments (QRAs) based on experiential understanding of the effect of uncertainty in relation to the programme out-turn cost (or anticipated final cost [AFC]) and schedule delivery will be performed at key stages of a programme. Certainty of outcome will become greater as the programme approaches its final stages. Probability management will store uncertainties as data in a model that is actionable, additive and auditable.

For this purpose, three-point estimating (as illustrated in Figure 4.5) can be used as a tool for estimating cost and schedule value and assessing overall risk.

The following areas will need to be considered during this process:

- Accuracy of data and assumptions
- Inconsistencies in data set
- Data errors

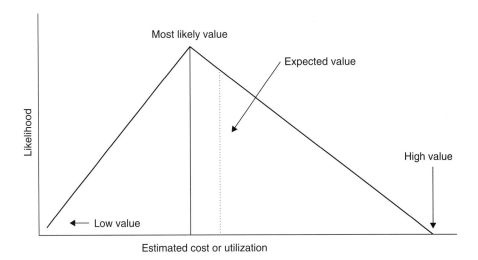

Figure 4.5 Three- point estimate triangle.

- Need for a wide range of perspectives
- Consideration to risk mitigation plan that can limit risk impact

Given the number of estimates and activities on a programme, the three-point estimate can sometimes be misleading. Another estimating technique is to enter variances of the probability distribution around the most likely estimate. This technique is based on an integrated holistic understanding and current knowledge of the programme's inherent risks. The estimation of uncertainty is illustrated in Figure 4.6.

A QRA undertaken on the programme will confirm the appropriate level of contingency (also known as 'overall risk pot') required to deliver the programme and shared between funding organisation, programme and projects. This includes the assessment of risks within individual projects in addition to cost and schedule risks across the programme.

The basis of the risk model should be an agreement of risk allocation between funders. The level of contingency proposed in the budget should represent those risks agreed to be under the influence and control of funders, programme and projects. The funders contingency relates to risk that does not sit with the programme and projects. The programme contingency relates to risks that do not sit within the individual projects and are not covered by project contingencies.

In summary, an s-curve will represent the likely exposure of the total risk across the programme (see Figure 4.7). Based on the agreed allocation of risk (at the 80th percentile),

	Classification	Uncertainty	Overrun
A	Routine, been done before	Low	0% to 2%
B	Routine, but possible difficulties	Medium to low	2% to 5%
C	Development, with little technical difficulty	Medium	5% to 10%
D	Development, but some technical difficulty	Medium high	10% to 15%
E	Significant effort, technical challenge	High	15% to 25%
F	No experience in this area	Very high	25% to 50%

*The % range for uncertainty will depend on the appetite of the organisation.

Figure 4.6 Estimation of uncertainty: illustrative example.

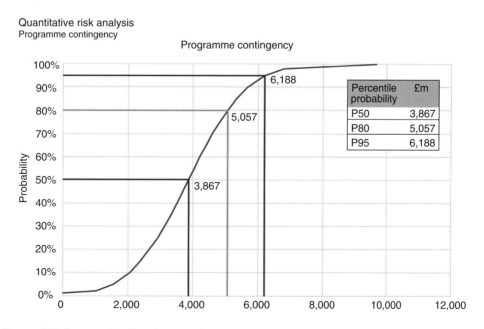

Figure 4.7 S-curve detailing the cumulative contingency requirement.

the contingency requirement, inclusive of VAT, will be estimated – that is, there is an 80% likelihood of not exceeding this contingency. The curve will also indicate that, from the analysis, an additional £xxxm would be required to secure a confidence at the 95% level.

The s-curve will indicate an upper limit that is significantly greater than the 80th percentile; to some extent this is influenced by the probabilistic model and the potential for uncontrollable acceleration costs or design risks. This suggests an increased maximum out-turn cost, albeit at a low level of probability. It is important that funders recognise increased exposure throughout the programme as it represents a substantially higher potential cost above the level at which the programme contingency is calculated.

There are other ways to represent risk and probability graphically as tornado diagrams or bar charts using Monte Carlo analysis and proprietary software for cost and time probability assessment.

It is apparent from programme delivery history that the risk of strategic misrepresentation remains high for complex programmes. Senior managers will need to consider the level of optimism biased (the tendency to overestimate the achievability of planned actions) for any programme. Benchmarking, due diligence, and historical local performance analysis will allow the programme to avoid blind spots and prepare the organisation for black or grey 'swan events' – low-probability high-impact risk (terrorism or major temporary design failure) or unknown unknowns (natural disaster).

4.4.4 Governance of programme management: steering for success

Governance defines the structure, roles and responsibilities to set objectives and report and monitor performance in order to make decisions and to steer a programme towards its anticipated destination.

The Organisation for Economic Co-operation and Development (OECD) defines governance as a set of relationships between an organisation's owners, its board, its management, and other stakeholders. This provides the structure through which the organisation's objectives are set and the means of attaining those objectives and of monitoring performance are determined.[3]

The APM (Association for Project Management) defines governance of programmes and projects as follows: governance of programme management concerns areas of corporate governance that are specifically related to programme and project activities. Effective governance of programme management ensures that an organisation's programme is aligned to the organisation's objectives, is delivered efficiently and is sustainable. Governance of programme management also supports the means by which the board, and other major project stakeholders, are provided with timely, relevant and reliable information.

The governance of programmes will cover the following areas:

- Leadership and sponsorship, clear and documented roles and responsibilities
- Strategic direction, a multidimensional roadmap to reach the final destination
- Programme management methodology, including a set of policies, procedures and processes to control programme trajectory against baseline
- Integrated assurance – transparency and disclosure to all stakeholders via operational (projects), functional (PMO, department) and compliance (risk and audit) lines of defences

---

[3] *OECD Principles of Corporate Governance* (2004), OECD, Paris. Available at: http://www.oecd.org.

Stage C: Definition

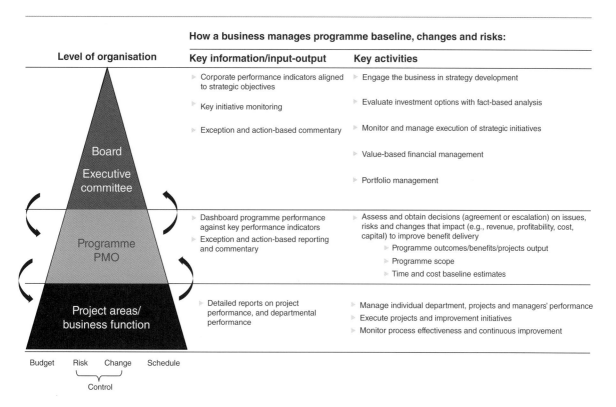

Figure 4.8 Change management, risk management and reporting.

The terms of reference for a programme board will typically cover the following:

- Requirement – why?
- Role – what?
- Composition – who?
- Conduct – when?
- Proposed delegated authority – how/how much? This is based on funding agreement and contingency allocation

Optimal governance, programme change control, risk management and reporting of an approved baseline scope will allow a business to manage risk, deliver value and drive programme and project decision-making while saving time and efforts. See Figure 4.8 and Figure 4.9.

Establishing and implementing a measurement and reporting system is a complex and evolving process based on the simple business principle that what can be measured can be managed. 'Too much data, not enough information' is a common complaint among business managers who have to keep track of the status of programmes while relying on information that is incomplete, out of date, inaccurate, late or simply irrelevant.

Reporting requirements will change as projects and programmes move from inception through their life cycles. It is important that the reporting system is designed to focus on metrics that matter at the relevant stage.

The critical success factors that underpin the operation of effective reporting are the following:

- Strategic objectives – Performance measures need to be aligned with the strategic objectives of the organisation. These objectives must be clearly communicated and understood by employees and external stakeholders

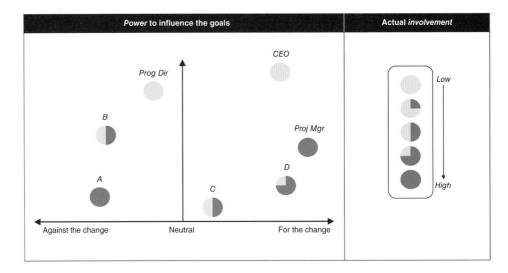

Figure 4.9 Ability to impact and commitment to the change.

- Baseline management – The baseline scope, programme timescales, resources and budget cost for the approved brief are clearly and consistently communicated to all parties

- Work breakdown structure – A WBS is designed to provide a common basis for linking the scope of work, estimates, budgets, schedules, earned value progress, performance and cost reports, based specifically on the customer's brief

- Progress management – As the programme progresses, the actual delivery times, resources and costs are recorded, analysed and compared to the baseline

- Change management – All changes from the original baseline are managed and incorporated into subsequent controlled copies of the original baseline called 'the current baseline'. The current baseline is progressed to reflect the current situation and subjected to analysis

- Reporting – All baseline and progress information is collected, analysed and reported in a simple yet robust manner based around schedule, cost and quality performance reports, clearly communicating the status of the delivery brief to all parties, including the client

The measures that matter for effective control should be simple and reliable.

Key performance indicator themes should be reported and agreed upon by each functional area or, in leading global organisations, across a portfolio of projects or a programme to deliver strategic objectives. The indicators are then aggregated into project, programme and executive board reports to assist in meeting the needs for information, control and governance. As the programme progresses through its life cycle, the quality and availability of key performance data available to the different functions will develop.

Performance measures should be based on the key performance indicators that are most commonly associated with the built environment and include key objectives such as cost, time and quality.

Organisations will also seek to measure and report other success factors that align with their strategic objectives, such as environmental sustainability and corporate social responsibility themes. The importance of these performance indicators and measures will change as the capital projects and programmes move through the various stages of their life cycle (investment planning, design, procurement, manufacturing, construction, commissioning and operations).

Once this process is in place and data and information are flowing more or less painlessly, it is then down to management to trust it and act on it, as there is often no worse decision than no decision at all – even a bad one.

This trust will undoubtedly be reinforced by a level of checks at all levels of the organisation, or integrated assurance.

4.4.5 Issues management

As outlined in Section 4.4.3, issues management is an ongoing programme management function, akin to risk management. The programme's approach to issues management must be clearly defined and endorsed by the PrgS, including the decision-making procedures.

The approach should include roles and responsibilities in relation to management of issues during the programme and may include the following:

- How issues will be flagged, passed up the hierarchy, recorded and assessed
- How issues will be monitored and controlled (both at project level and programme level)
- Who will generate the report and when, and who will receive the reports and determine action
- What will be the regular review points
- What will be the escalation procedures

Initial issues will have previously been recorded as part of the completion of the brief and the business case – the initial registers can be turned into the live issues register with assigned ownerships as the programme progresses.

4.4.6 Time scheduling

In its simplest form, a schedule is a listing of activities and events organised by time. In its more complex form, the process examines all programme activities and their relationships (interdependencies) to each other in terms of realistic constraints of time, budget and people, that is, resources. In programme management practice, the schedule is a powerful planning, control and communications tool that, when properly executed, supports time and cost estimates, opens communication among personnel involved in programme activities and establishes a commitment to programme activities.

However, it is not always practical or realistic to prepare a time schedule for programmes – in many instances there may be additional constraints (e.g. output delivery dates may be set by parameters beyond the control of the delivery team or a target may be set for commencement of benefits realisation for certain outcomes); it is often the case that, especially for programmes with intangible deliverables, while a realistic time schedule can be prepared for individual projects and activities, the overall programme time scheduling remains a bit of an unknown until the commencement of the implementation phase. However, almost all business cases will allocate a certain time for any programme even if only for funding purposes, and it is the responsibility of the PrgM to ensure that a detailed programme schedule is prepared at the definition phase identifying each individual component and that it is regularly reviewed for any variance. The PrgSB will be made aware of the variances through the programme highlight report and should decide on the appropriate course of action.

Role	Responsibility
Programme manager	✓ Ensure that appropriate financial controls are defined and operate for the programme ✓ Ensure that programme operates within the allocated budget ✓ Provide timely reports to PrgS and PrgSB of any significant variances from the approved budget or changes and or issues that may result in a variance
Programme finance manager	✓ Provide advice and guidance on the proper financial management of the programme and each project or activity ✓ Oversee compliance and governance in terms of financial management requirements ✓ Maintain up-to-date information on budgets, forecasts and expenditures ✓ Provide regular reports to PrgM, PrgS and PrgSB
Project managers	✓ Ensure that any overarching financial management and governance requirements and enshrined within project documentation ✓ Ensure that projects are delivered within the allocated budget ✓ Provide regular reports to PrgM and project board

Figure 4.10 Financial management roles and responsibilities.

4.4.7 Financial management

Business today needs finance officers who understand how engineering and production processes are really managed to help with establishing:

- clarity in budget and contingency allocation between funder, programme and projects

- understanding of reporting requirements at corporate and programme level in order to prevent misalignment, inconsistencies and misinterpretation

Difficulties in aligning and reporting for business and programme purposes are not uncommon. Financial reporting normally spans a fiscal year, whereas programmes, due to their nature, may span a number of years. Additionally, financial and programme professionals do not necessarily talk the same language or have the same reporting needs.

Communication is the key aspect of prudent financial management. Opacity in budget allocation leads managers to make assumptions and decide on the basis of their current knowledge over time. Particularly on major programmes, it is of vital importance that meaningful and timely information is generated from the vast amount of data that would be available. Figure 4.10 sets out the individual roles of the key personnel in context of programme management at this stage.

Funding arrangements

Once funding arrangements, share of contribution, delegated authority and contingency allocation is apportioned and agreed upon to form a fund, a budget developed in parallel can be managed and its change controlled through the programme gates.

Budget, including contingency, can be represented by a WBS element using a tabular format or a bar chart as seen in Figure 4.11.

It is essential to capture baseline assumptions as part of a budget to allow for all parties to understand the associated risks. Assumptions will typically include: scope, design, construction, schedule, cost, inflation, currency, commercial, procurement, land and property, legal framework, operations, and asset management.

Changes and contingency management related to scope (schedule, cost and quality), including design development, valuation (value engineering), new scope, sequencing or acceleration cost (also known as de-risking in some programmes), cost savings or

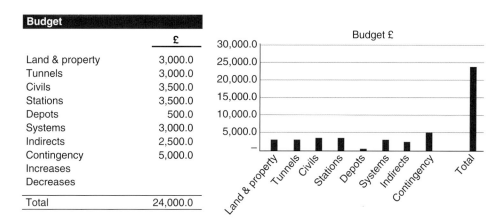

Figure 4.11 Programme budget for transport programme (example).

procurement and delivery model adoption (for example engineering procurement and construction as opposed to in-house delivery) can be identified and communicated using a waterfall or bridge diagram.

Waterfall charts[4] are commonly used in business to show how a value changes from one state to another through a series of intermediate changes. Waterfall charts are often called bridge charts (particularly in financial jargon) because a waterfall chart shows a bridge connecting its endpoints.

4.4.8 Cost management

Traditional cost and performance measurement systems often fail to distinguish between cost incurred and physical progress made. Under these systems, it is difficult to analyse separately cost variances and schedule variances. One answer to this is earned value management (EVM).

There are two key primary objectives of an earned value system: (i) to provide programme project managers with a reporting system that gives them better control over cost and schedule management and (ii) to provide customers a better picture of the status of work. See Appendix T5 for a Monthly Programme Report template.

Cost performance indicators (CPIs) and schedule performance indicators (SPIs) are gaining popularity as performance measurement systems in the construction industry and are acknowledged to make a positive contribution to the planning and overall control of a major construction project or programme. CPI = 1 and SPI = 1 means the programme is on track.

CPI and SPI are the two principal components of EVM (see Figure 4.12) and allow project and programme to:

- Measure programme and project progress
- Forecast completion date and final cost
- Identify schedule and budget variances along the way

There are several secondary objectives for measuring CPI and SPI on projects across a major programme. These include:

- Providing a timely 'early warning' signal for prompt corrective action
- Comparing the amount of work performed during a period of time indicating whether the project is behind or ahead of schedule

---

[4] A waterfall chart is a form of presenting the data in a visual manner that helps make it easier to understand the cumulative effects of positive and negative values in sequence.

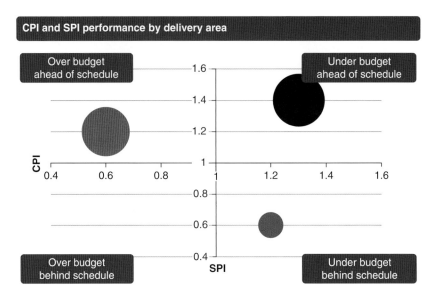

Figure 4.12 Delivery/project performance – programme EVM summary.

- Comparing the budgeted cost of work performed with actual cost, indicating whether the programme is over budget or under budget
- Encouraging contractors to use effective internal cost and schedule management systems

By integrating these measurements, EVM aims to provide consistent indicators to evaluate and compare the progress of programmes and construction projects. EVM compares the planned amount of work with what has actually been completed to determine if cost, schedule and work accomplished are progressing as planned. Work is 'earned' or credited as it is completed.

CPI and SPI are not absolute measures of progress and should not be viewed in isolation. The indicators should be viewed in conjunction with other progress reports and performance measures such as monthly milestone and variance reports. The information is processed and published in progress reports on a monthly basis and, in conjunction with other project progress information, can be used by both the client and supplier(s) to monitor performance and encourage a greater degree of control and proactive decision-making.

To make a positive contribution, the measurement system is dependent on two key factors:

- Implementation and integration of a collective and accurate breakdown of the construction works that capture the planned schedule (time) and budget costs throughout the lifetime of the project or programme. This is referred to as the cost loaded schedule
- Implementation of a robust and rigorous system to regularly capture actual time and cost data in order to compare it with that planned

Programmes will also report an annual spend forecast (Figure 4.13) or fiscal year performance and provide long-term projections (Figure 4.14) based against the approved baseline. This should be aligned and coordinated with financial reporting.

Although a source of many frustrations, integration, as illustrated in Figure 4.15, is key to programme and financial reporting alignment and will drive accurate reporting and inform decision-making.

One method is to set up the main accounting systems to track cost and billings by WBS line item. Most ERP (Enterprise Resource Planning) packages will track inventory movements by item type or by project code.

Stage C: Definition

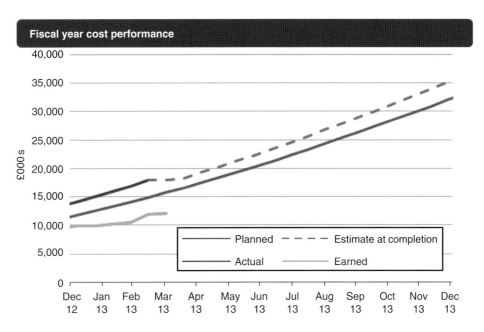

Figure 4.13 Programme fiscal year performance (annual spend forecast).

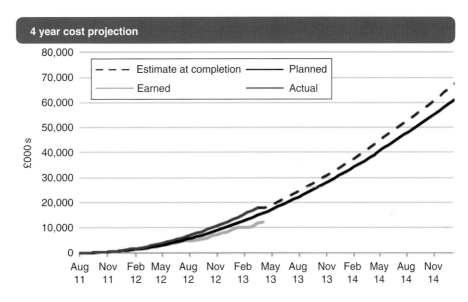

Figure 4.14 Four-year programme cost projection.

Project- and programme-oriented businesses need to be able to relate the use of material and labour to one key 'manufacture' or 'procurement' item number at a WBS level, allowing:

- Meaningful comparison between the estimated or budgeted cost of work performed (BCWP) and the actual cost of work performed (ACWP)
- Real-time update of the estimates or standards for costs to complete against actual cost based on incurred costs

Best practice strategies for improved cost control include:

- Cost management planning
- Supplier involvement
- Value engineering
- Cost reporting

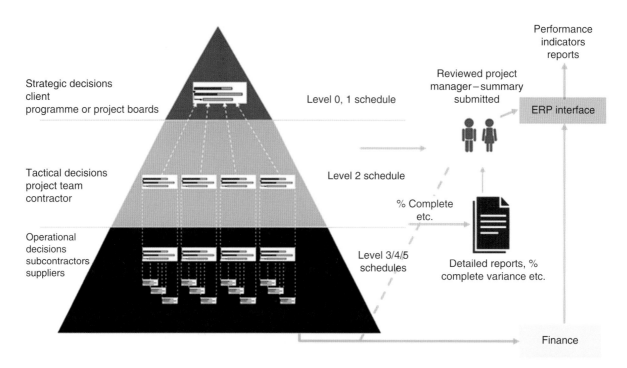

Figure 4.15 Reporting integration.

- Programme cost change management
- Cost forecasting
- Risk reporting

Financial management

Finance will control the programme expenditure at year end and monitor actual against forecast (AFC) expenses and budget. Profit and loss and balance sheets will be also compiled for business-reporting purposes and will include programme cost to date and any other additional costs outside programme control. (See Figure 4.16.)

4.4.9 Change control

As the saying goes, the only constant is change. At the definition phase, it is critical to ensure that a robust change control and management procedure is designed to enable formal identification, evaluation and decision-making required for accepting or rejecting changes to aspects of the programme.

Procedures should be put in place to manage changes, both at programme level and project level. Changes to scope, PDP, business case, benefits and budget will affect the overall programme and must be managed at the programme level.

Programme level changes must be signed off by the PrgSB and be managed by PrgM.

Typically, programme highlight reports should contain a change register.

4.4.10 Information management

Process and systems to be used for managing and storing information include:

- Programme level BIM system and process
- Responsibilities for issuing, distributing and maintaining information
- Archiving and retrieval protocols

Information management: sharing knowledge

At the heart of a programme is information and how it is communicated. Information has to flow within a programme in a way that will allow management, project managers and project teams to achieve business objectives while transparently keeping all stakeholders informed.

Communication is the process of transferring information from one source to another. Effective communication is a two-way process in which technical information, analysis, commentary and comments are conveyed through reports, speech and other mediums to a specific audience so that the intended audience will be informed, perform an action, or reach a decision based on this information.

The fundamental purpose of information management is to communicate effectively and with consistency relevant and accurate data to project stakeholders at all stages of the programme life cycle and beyond using effective information systems and models – BIM. The provision of good quality, timely information is an essential deliverable

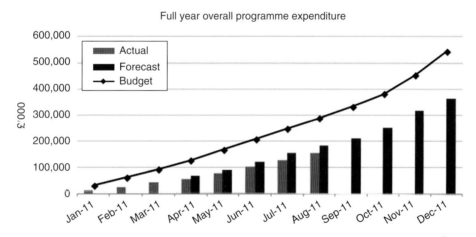

Figure 4.16 Full year programme expenditure example.

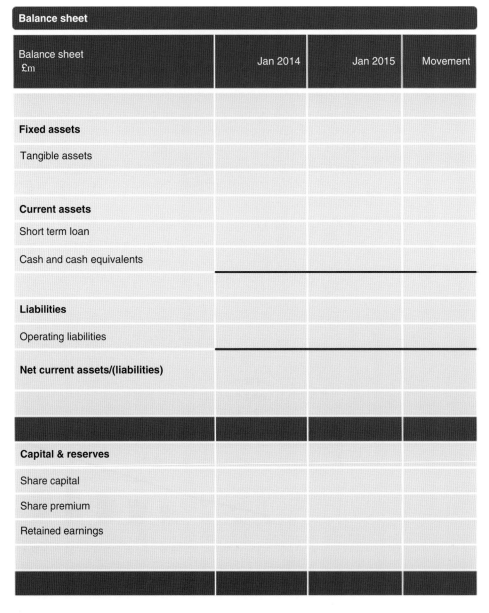

Figure 4.16 (Continued)

for any programme. Systems and procedures should be defined to enable effective management of information through the life cycle of the programme from inception to the delivery of operational assets. An effective information management system should facilitate direct relationships with the relevant information enabling storage and 'retrievability' of the following:

- Document ID
- Document location
- Document classification
- Document function
- Document status (cost and time)
- Document criticality and confidentiality

Integration

Integration is the combination and co-ordination of multiple sources of related information so that sources work together and form an 'integrated' whole with the ability to provide a snapshot of progress status at any chosen reporting milestone. The integration process will allow for interdependencies and design, procurement, construction, commissioning, departmental, organisational and supply chain interface issues to be identified and addressed from a single source. Key information (variance, exception) will be captured, assessed and forwarded to the relevant authority for a decision to be made and implemented.

The level of integration will vary from loose to tight integration depending on the current programme stage, agreement of a common purpose, set of objectives and level of process and systems standardisation. Programmes are usually carried out by a project team under the overall direction and supervision of a PrgM. In practice, there will be many variants of this structure, depending on the nature of the programme, the contractual arrangements, type of project management involved (external or in-house) and client's requirements.

As the programme evolves, it is essential to strive for an integrated information repository to manage increasingly detailed information. All integrated programme assets or documents should be coded and linked to a programme database. In the context of a physical asset programme, BIM is a powerful tool that will allow a programme to be integrated in a multidimensional model created for management purposes.

In summary, programme management systems need good quality sources of information on which to base their decision-making and manage operations post-delivery. Management reports, project summaries, reporting systems and data will include the following:

- Programme and project costs incurred to date and current estimate also known as out-turn cost (i.e. how much will it all cost in the end?)
- Cash flow status on programme and project and estimated cash flow forecast for the remaining period of programme
- Earned value measures of schedule variance and cost variances for project work packages and overall programme
- Status of risk management actions in relation to the database of programme risk registers
- Estimates of risk and uncertainty remaining in the key elements of the programme

4.4.11 Communication/stakeholder management

The process developed for communicating with the programme team and stakeholders may include:

- Responsibilities for generating and for distributing communications
- Nature of communications to be issued
- Methods of communications to be utilised
- Stakeholder analysis
- Frequency of communications

Stakeholder identification and management

A stakeholder management plan aligned to business and programme objectives can be designed in six steps to communicate effectively to the programme audience and gain support along the way.

1. Define, restate and clarify objectives (link back to strategic change objectives).
2. Identify stakeholders, internally and externally.
3. Analyse and map stakeholders by proximity into top tiers, second tiers and so on.
4. Plan management of stakeholder communications and reporting.
5. Identify the right channels to effectively engage with your stakeholders.
6. Monitor progress in stakeholder engagement, persuasion and advocacy levels against a programme timeline.

Defining stakeholder segments starts during the strategy and programme definition phases.

Programmes and organisations need to be able to identify their stakeholders and assess the level of impact they will have in meeting the objectives and outcomes of the programme. A first stage for this process will be to create a stakeholder proximity map. This map will lay out different tiers of stakeholders according to their potential to affect the reputation or the progress of the project or organisation, with the programme at the centre.

Stakeholders will be identified, captured and mapped for proximity to the programme in terms of direct impact or interest, influence and control in the programme landscape, based on the following questions.

- Who makes up the most important tiers of stakeholders for this programme?
- How great is their impact on the programme progress or reputation?
- How great is their influence on other important stakeholders, for example those who control resources?
- What is their level of engagement in the present time: are they in favour, against or indecisive?
- How easily can they be influenced, persuaded and brought on board as long-term advocates and supporters, especially in times of issues and crises?

Once the stakeholder map and landscape is understood, a tracking tool articulating what reengagement activities are required for each stakeholder will allow tracking progress. The tracker will identify the stakeholder relationship owner in the programme – a person accountable for managing and directing the stakeholder relationship.

Making the link with the communication team

It is a common experience in a professional's life to have the urge and necessity to communicate a message that, at that time, was thought to be right, needed to be communicated without delay and needed to be communicated to a group of colleagues and line managers. It is also fairly common, with the benefit of the hindsight, to think about an e-mail that has been sent and should not have been. If this communication is scaled up to a very large organisation and programme and the way this message communicated the message is considered, particularly in times of critical decisions, the importance of clear, efficient, effective and appropriate communication can perhaps be put in some context.

In simple terms, communication is the process of transferring information and knowledge from one source to another.

Effective communication of the relevant information is a two-way process in which data, evidence, analysis, commentary and feedback are conveyed through reports, speeches and other media to a targeted audience to perform an action or reach a decision based on this information

Once the stakeholder audiences are defined, it is paramount to define a set of key messages and positioning statements to address these audiences in their questions, concerns, and other key requirements. A series of pre-defined lines to adopt should also be prepared and ready to use, especially in cases of crises or issues. It is important that messages are continuously updated and aligned with the programme's variables or changing milestones.

Communications strategy and management

A good communication strategy and plan will ensure that the programme audience is informed in a targeted and timely manner.

Based on the stakeholder analysis, the communication plan will list the messages and interventions to be communicated to which audience group, when, how (medium), why and from whom.

Communication mediums can be categorised as high or low impact and commitment building and will include:

- One-to-one coaching or small-group workshop
- Intranet, social network, knowledge transfer
- Large group, conference
- User guides, reminders, online help

Key communication principles include the following attributes:

- Benefit led
- Messages that pinpoint the benefits that are relevant to stakeholders
- Timely, accurate, up to date, regular and consistent
- Appropriate in language and media for stakeholders and staff
- Whenever appropriate, use existing channels
- Encourage two-way dialogue and feedback

4.4.12 Quality management

The method and responsibility for assessing the 'fitness for purpose' of the project outputs, outcomes and programme deliverables will need to be defined. This might be through specifying the acceptance criteria and/or the performance criteria. For programmes with tangible deliverables, for example, a new facility, it is relatively easier to define the quality acceptance criteria; where programmes include intangible deliverables, it may be necessary to rely on secondary assessment methods to quantify acceptance or performance criteria.

It is advisable to introduce periodic reviews of the quality management procedures at key stages of the programme life cycle to test the adequacy and the effectiveness and enable any changes or amendments as necessary.

4.4.13 Procurement and commercial management

Procurement is the process by which the resources (goods and services) required by a project are acquired. It includes the development of the procurement strategy, preparation of contracts, selection and acquisition of suppliers, and management of the contracts.

The benefits of contracts and procurement are that:

- A project in which procurement is aligned with the programme deliverables is more likely to meet its objectives
- Procurement often represents a major portion of project spending and hence needs careful consideration to ensure value for money is realised
- It ensures that all parties involved in the project are legally protected

Effective procurement and contract management is core to a successful programme. The procurement policy, procedures and processes should be established to guide the procurement of a range of contracts and projects.

It is important for companies to develop purchasing policies and procedures that include the following:

- Supply chain management
- Strategic partnerships
- Market knowledge
- Cost reduction
- Sustainability

Based on the above, managers will develop a 'contract packaging' approach which will break down the entire programme of work for delivery into suitable elements for procurement. This will be refined in discussion with the potential suppliers and contractors. To meet the programme objectives, on time and in the given budget, often managers will procure a large number of contracts and projects of varying size and value.

A robust, fair and transparent approach to procuring, managing and monitoring these agreements is essential, as is carefully managing the risks associated with the overall programme and related projects. A programme will often have to operate in the procurement framework set out by international procurement legislation and national regulations.

For major programmes, it is also recommended to give advance notice of contracts through a 'future opportunities' website – the equivalent of market building for supply chain members – and adopt e-tendering practices.

To operate effectively in the marketplace, programmes will use standard procurement documentation and contracts such New Engineering Contract (NEC)/Fédération Internationale Des Ingénieurs-Conseils (FIDIC) and CIOB's.

Managers will award contracts according to broader objectives and values. Ahead of every contract competition, programmes will develop a 'balanced scorecard' of selection and award criteria that will inform companies of the commercial and technical factors their bids will be scored against. This should include time, quality, safety and security, equalities and inclusion, sustainability and legacy.

Appropriate criteria will be added and different weights applied to each criterion on a contract-by-contract basis, according to the nature of the goods, services or works being procured and best practice guidance from the Office of Government Commerce. To avoid costly and time-consuming disputes, the programme should seek to make clear agreements with all parties while agreeing and managing contracts.

Best practice suggestions to manage the risks in this area include:

- Developing a procurement plan
- Establishing vendor selection criteria
- Having an overall buying strategy for negotiation
- Establishing quality-control procedures and a set of performance measures.

Key performance indicators to measure the performance of the procurement or buying processes include:

- Unit price for comparison
- Total cost of supply
- Number of suppliers
- Average purchase order size
- Number of contract out, placed or closed
- Value of contract placed
- Satisfaction indices

Figure 4.17 illustrates an example of progress of procurement and commercial management presented graphically.

Resources management: building a team

Managers need to 'design' their organisations and programmes to embrace change and deliver programmes effectively.

$$Talent \approx f(motivation, competence)$$

The following criteria will need to be considered in order to design and agree on resource profiles that are fit for their purposes and align with the delivery model so that the right level of resources are allocated to a programme at the right time:

- Work breakdown structure (WBS) – a key tool for efficient and effective resource allocation to create resource profiles for delivery
- Scope/size – Bill of quantities (BoQ) and work packages; scope definition, including a hierarchical list of materials, components, assemblies and sub-assemblies required to make a manufactured product
- Organisation breakdown structure (OBS) and organisational policies: Structure and a clear statement of company policy on key areas such as staffing, subcontractors' issues and procurement policies
- Benchmarks and historical information. Information from past records
- Team experience and clear roles and responsibilities (using a RACI model) within a programme and down the supply chain with appropriate level of "man marking and risk transfer"

The above criteria will depend on the delivery model adopted, but leading PrgMs focus on building an intelligent client organisation to lead delivery and maintain the right

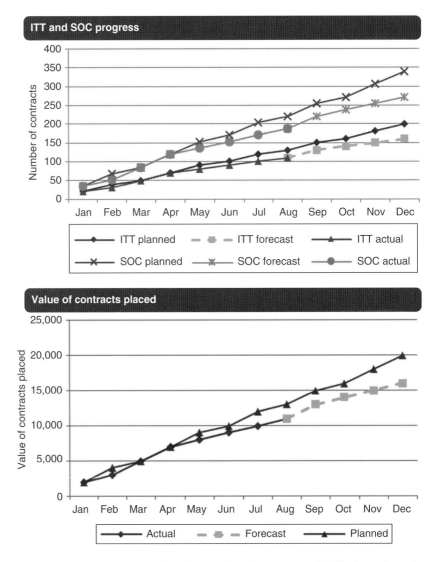

Figure 4.17 Invitation to tender (ITT) and signed outline contract (SOC) plus value of contract placed.

level of control. This is a key factor in successful cost management. It needs to be established at start of programme so that appropriate resources are allocated. Where necessary, certain elements may then be outsourced, subcontracted or bought according to a build or buy model.

4.4.14 Health and safety management

Programmes will need to formulate health and safety management policies and procedures both at the project level and the programme level. It is the responsibility of the PrgM to ensure that at the definition phase the health and safety goals, requirements and procedures are set and that the processes are put in place to transfer these to the project levels as necessary. In some instances, specific key performance indicators or targets may be set to measure performances both at the project level and the overall programme level. It is advisable to also set specific review stages to assess whether the health and safety management policies and procedures are performing as desired across the programme.

4.4.15 Sustainability/environmental management

It is often said that sustainability is essentially not a methodology but a dimension of thinking. The perception of sustainability is shaped by people's values, behaviour, attitude, ethos and interaction with the wider environment. Programme management is concerned more with outcomes than outputs. With emergent input in a changing environment, a PrgM makes use of current information to identify options for comparison and decision. Once the decision is made, the PrgM will take over the project(s) where sustainability is a consideration. In other words, for sustainability in programme management, the PrgM emphasises the learning loop leading to the preferred options: the PrgM has to assess the suggested programme options in the dimensions of economic sustainability, environmental sustainability, and social sustainability before recommending the options. In most cases, the organisation commissioning the programme will have sustainability and environmental management policies at a wider level or, indeed, may have certain aspirations and targets specific to the programme. The PrgM must ensure that the goals and targets, as necessary, are defined and translated to procedural requirements at the project level.

5 Stage D: Implementation

> - Are the management processes to control, direct and manage the programme being used during the implementation phase?
> - Is the programme actively managing the programme delivery plan, benefits and business case and engaging with stakeholders?
> - Is the programme managing the transition from the old to the new ways of working?

5.1 Purpose of stage

The purpose of this stage is the initiation and execution of the various projects comprising the programme, including assessing performance of individual projects, managing the interfaces between projects, monitoring benefits realisation, managing financial expenditure and managing the introduction of any changes to the programme.

5.2 Stage outline

In this stage the physical activities of the programme are executed through a number of projects (see Figure 5.1). The focus of the programme management team is on ensuring the successful implementation of each individual project in accordance with the planned sequence and the delivery of the required outputs. At the commencement of the stage, the following tasks need to be actioned.

Initiate projects

- Appoint the initial project management teams
- Issue terms of reference and brief the project teams
- Liaise (the project management office [PMO]) with the governing structures to establish processes and procedures

Performance monitoring and control

- Implement the management processes and procedures to monitor and control the programme in the areas of:
    - Stakeholder management
    - Communications

Code of Practice for Programme Management in the Built Environment, First Edition. The Chartered Institute of Building.
© 2016 John Wiley & Sons, Ltd. Published 2016 by John Wiley & Sons, Ltd.

Code of Practice for Programme Management in the Built Environment

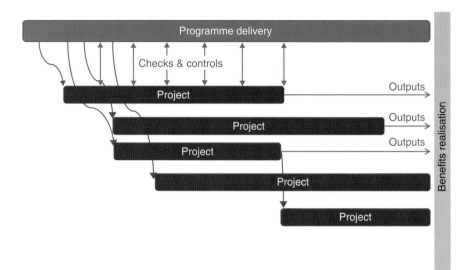

Figure 5.1 Stage D: Implementation.

- Financial management
- Benefits realisation
- Governance
    - Risks/issues
    - Time
    - Costs
    - Performance
- Quality management
- Monitor progress of implementation against the programme delivery plan (PDP).

Reporting

- Report regularly to the programme sponsor (PrgS)/programme sponsor's board (PrgSB) on the progress of programme delivery
- Raise any critical issues requiring consideration by the PrgSB

Projects closure

- Manage the formal closure of projects, validate project outputs and take ownership of outputs
- Carry out end-of-project reviews and compare actual achievement with the required outcomes and benefit realisation
- Carry out lessons-learned review and highlight any aspects that will benefit other projects

5.3 Stage organisation structure

5.3.1 Stage structure and relationships

The implementation stage marks the maximum effort and involvement of the full programme management team as there is a collective effort to deliver the outcomes of

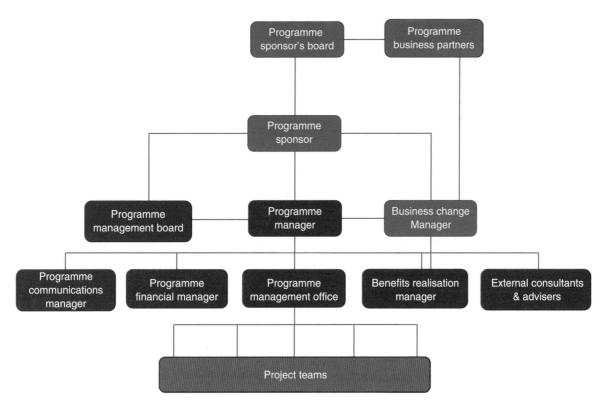

Figure 5.2 Stage D: Implementation – organization structure.

the programme in accordance with the proposals set out in the PDP. This effort is headed by the programme manager (PrgM) who, closely supported by the PMO, initiates, monitors and closes the projects which make up the programme. The business case manager, together with the benefits realisation manager, ensures that the required benefits are being progressively realised and that the new capabilities or facilities are being effectively introduced into the client's organisation. (See Figure 5.2.)

5.3.2 Stage roles of key participants

Programme sponsor's board

The PrgSB continues its role as adviser and approver, fulfilling the following duties:

- Reviewing regular reports on the programme's progress
- Providing overview of the progressive realisation of benefits
- Resolving any issues raised by the PrgS
- Providing any approvals or decisions on matters raised by the PrgS
- Providing overview of the introduction of any client-generated changes

Programme sponsor

Continuing to be the interface between the programme and the client's organisation, the PrgS maintains a check on the programme's overall progress and keeps the client advised of this progress. Principal tasks include:

- Regularly reviewing progress towards the programme's objectives
- Regularly monitoring performance of the project management team
- Referring any major issues to the PrgSB

- Obtaining any approvals/decisions required from the PrgSB
- Verifying the PrgM's recommendation for appointment of project managers
- Dealing with any client requirements for changes to programme scope, objectives or deliverables
- Determining, in conjunction with the PrgM, any additional projects required to maintain (or enhance) the planned deliverables
- Depending on the nature and complexity of the programme, it may be necessary for the PrgS and/or PrgM to initiate either internal or external audits of the programme or projects

Business change manager

The business change manager (BCM) is focused on ensuring that the programme is achieving its planned objectives by:

- Monitoring achievement of deliverables against the plan
- Managing the integration of the changes into the client's organisation
- Keeping aware of any changing external circumstances that may affect the outcomes of the programme and impact benefits

Benefits realisation manager

Concerned with ensuring that the programme is delivering benefits in accordance with the PDP, the benefits realisation manager has the following responsibilities:

- Monitoring benefits being delivered by projects against the criteria set out in the benefit profiles
- Reviewing benefit realisation against the key success criteria
- Assessing the effectiveness of the benefits being realised
- Reporting any variances in benefits being realised to the BCM/PrgM

Programme manager

With full responsibility for the implementation of the programme, the PrgM has a wide range of tasks, including the following:

- Initiating projects in accordance with the PDP
- Selecting and appoint project managers
- Approving the selection of project teams and consultants
- Monitoring the progress on individual projects
- Monitoring the performance of project teams
- Monitoring the quality of outputs from projects
- Approving the closure of projects
- Liaising with the BCM on transition arrangements at the closure of each project
- Resolving any issues on projects
- Raising any major issues with the programme management board

- Reviewing regularly progress of the programme in terms of time, cost, risks, deliverables, etc. with the PMO
- Reviewing regularly stakeholder issues with the stakeholder/communications manager
- Reviewing regularly financial issues with the programme financial manager
- Reviewing regularly benefits realisation with the BCM/benefits realisation manager
- Reviewing regularly health and safety issues with the health and safety manager
- Reviewing regularly sustainability issues with the sustainability manager
- Reporting regularly progress of the programme to the PrgS
- Referring any critical issues to the PrgS
- Managing the introduction of any changes instructed by the PrgS

Programme financial manager

Throughout the implementation stage, the programme financial manager continues to manage the financial aspects of the programme. The tasks of the PrgFM will include:

- In conjunction with the PrgM, monitors and reports to PrgS and client on programme expenditure and the predicted final cost
- Liaises with funder(s) on the release of monies
- Manages risk and contingency allowances
- Reviews the programme's financial performance against the business case
- Monitors the cost implication of instructed client changes
- Reports any significant issues/problems to the PrgM

Programme management board

The PrgMB continues to provide advice and support to the PrgM during programme implementation. The programme management board:

- Reviews the programme's progress regularly
- Resolves any issues that are impacting progress
- Refers any significant issues to the PrgSB for advice or resolution

Stakeholder/communications manager

The stakeholder/communications manager continues to ensure interested stakeholders are identified and that there is a mechanism for maintaining their engagement in the objectives of the programme by doing the following:

- Reviewing regularly the stakeholder analysis
- Maintaining regular communication with stakeholders
- Maintaining regular communication with the client organisation
- Maintaining the activities identified in the public relations plan
- Reporting any significant issues/problems to the PrgM

Programme management office

The PMO is responsible for providing the technical functions that allow the programme to be managed and controlled, including the following tasks:

- Ongoing monitoring of progress on programme and projects
- Ensuring projects are being managed in accordance with the processes and procedures defined in the PDP
- Preparing regular reports on progress for programme management board and PrgS
- Maintaining change control process
- Maintaining issues log and highlighting major issues to PrgM

Within the PMO, functions with specific responsibilities include:

- Head of PMO: Maintains the ongoing operation of the PMO
- Scheduling manager: Monitors progress on the programme and all projects
- Cost manager: Maintains cost monitoring and control on the programme and projects
- Risk manager: Maintains the risk management process
- Document/information manager: Maintains the programme's document and information system
- Health and safety manager: Ensures the programme and projects comply with all relevant health and safety regulations and legalisation
- Sustainability manager: Ensures the programme and projects comply with all relevant sustainability regulations and legalisation and monitors progress against sustainability targets
- Administrator(s): Provides support to the PMO

Project management structures

Each project that is carried out as part of the programme will need its own management and technical teams responsible for the execution of the project. These teams are likely to include the following:

- Project managers
- Project support office(PSO)
    - Head of PSO
    - Scheduling manager
    - Cost manager
    - Risk manager
    - Document/information manager
    - Administrators
- Consultants and contractors
    - Design team
    - Contractor
    - Suppliers
    - Health and safety

5.4 Programme management practices

At this phase of the programme cycle, the procedural and management activities as defined during the earlier stages will be put into use to monitor and control the implementation of the outputs and outcomes.

It is the responsibility of the PrgM to ensure that resources across the programme are managed effectively. This will include the utilisation of internal and external resources and resources that may be shared across projects and programmes. Furthermore, the enabler technology and tools as identified during earlier stages to support the implementation should also be made available. It is often advisable, particularly for large and complex programmes, to have core teams co-located or at least undergo team-building exercises to develop effective relationships and engender trust.

In addition to the core programme management activities identified later, the following activities, which are initiated during the early stages of the programme, need ongoing attention during the implementation phase to ensure desired delivery of the programme outcomes and benefits:

- PDP
- Benefits management
- Stakeholder management
- Business case management
- Transition management

Programme delivery plan

Managing the PDP during delivery is vital to ensure that as the programme progresses, projects remain integrated with each other and the programme. In practice, this means regularly reviewing the PDP with the PrgM, BCM or their representatives and the project managers. It is important to give project managers the opportunity to report progress and BCM the opportunity to report operational issues. One way of doing this might be by undertaking PDP review workshops.

Holding a workshop with all project managers and change managers present will ensure that the knock-on effects of changes or ideas in one project or business area are explored with respect to the PDP in its entirety.

The regularity of PDP workshops will depend on what has been agreed on during the definition phase, but as a minimum should be undertaken as each step change in the programme is implemented.

The results of the review will usually mean that the PDP has to be updated and signed off before the next step change can be planned.

Benefits management

Benefits need to be regularly reviewed during the implementation stage as:

- Unforeseen issues and risks arise that require a change of plan
- Strategic priorities change in response to external circumstances
- Assumptions, and/or constraints, that were used to determine potential benefits are discovered to be inaccurate.

The method of review has been determined during the Definition Stage.

The benefit profiles will need to be updated and the benefit realisation activities in the PDP amended to reflect any changes.

Business case management

The business case is the basis upon which the implementation phase of the programme is commissioned. As delivery progresses, changes outside the control of the programme may mean that the assumptions and estimates upon which the business case was made change.

The business case should be kept up-to-date in response to any changes. It is the responsibility of the PrgSB during implementation to ensure that the business case is updated and, if the case is fundamentally affected, to explore changes to the PDP and benefits in order to maintain the case.

In the scenario where the business case is fundamentally no longer viable, it may be necessary to consider the termination of the programme.

Conversely, it is also possible that changed circumstances may lead to further strengthening of the business case. In practical terms, this will mean looking for opportunities to increase the benefits and to reduce the costs. In some cases, increasing the cost may facilitate greater benefits. However, this should be balanced against making too many changes that lead to unproductive use of resources or increased risk of not achieving the desired outcomes.

5.4.1 Performance monitoring, control and reporting

The amount and level of detail required for progress reporting varies widely from organisation to organisation and programme to programme. The approach to be taken, including the documentation to be used, was defined during the definition phase.

It is important that the information provided is complete, timely and accurate so that appropriate decisions can be made by the PrgSB. These decisions include the following:

- Decisions taken with respect to changes
- Review of outcomes, including any further actions necessary
- Review of realisation of benefits, including any further actions necessary
- Review of actual progress against planned progress and dealing with variance
- Review of the financial position for the programme and dealing with variance

Typically, PrgM will produce a programme highlight report which will form the basis of the review. An indicative template of this report is shown in Appendix T6.

5.4.2 Risk and issue management

Risk and issue management are ongoing programme management functions. The programme's approach to risk and issue management will follow the process defined during the definition phase.

Risk and issues registers should be regularly maintained and monitored. The actions agreed either to mitigate the risk or resolve the issue should be included in the overall PDP or monitored alongside the PDP.

It is important to ensure that, where necessary, risks which look as if they are going to materialise and issues which cannot be resolved are escalated. Who to escalate to and when will be documented in the approach defined during the definition phase.

Risks and issues are often recorded in one or more closely related documents. Whichever method is used, it is important to recognise that managing risks and issues are two separate processes. Issues may arise that need to be resolved before formal risk and issue review meetings. It's important that arrangements are put in place to immediately deal with issues as they arise. It is also possible that some issues may have an influence on the fundamental business case assumptions, in which case the PrgM should ensure that a considered resolution of the issue is undertaken.

5.4.3 Financial management

The financial status of the programme should be monitored in accordance with the process defined during the definition phase. This monitoring will include:

- Ensuring that appropriate financial controls are in place for the programme

- Ensuring that the projects and overall programme are being managed within budget

- Providing timely reports to the PrgSB on financial status and highlighting any significant variance from the budget or changes that might incur unplanned for additional expenditure (this can be included in the programme highlight report – see Appendix T6)

- Ensuring that the financial processes in place for the programme comply with corporate standards and regulations

- Maintaining up-to-date documentation on the financial status of the programme and the projects that are helping to deliver the overall programme

5.4.4 Change management

Change control is the formal identification, evaluation and decision-making required for accepting or rejecting changes to aspects of a programme. Management of change during the implementation phase should follow the approach defined during the definition phase.

The principle is usually that if a change can be managed at a project level, then this should be the preferred way of managing a change.

Changes to the PDP, business case, benefits and budget will affect the overall programme and should be evaluated and approved at the level of the PrgSB.

The following type of information and signed-off documents should typically be subject to change control:

- Core programme documents

- Programme governance

- Information with data protection issues

- Commercial and human relations sensitive information

- Programme communications

- Project-related information

A change can have a positive as well as a negative influence on the business case, and unless managed appropriately, it may ultimately lead to detrimental effects towards the realisation of the benefits envisaged.

5.4.5 Information management

Appropriate information management arrangements should be put in place in accordance with the agreed approach defined during the definition phase. This will ensure that information is available when it needs to be, is current, and is compliant with the quality standards set for the programme.

5.4.6 Stakeholder/communications management

During the delivery phase, it is essential to continue with planned engagement and communication activities with stakeholders as well as respond to unforeseen events.

The stakeholder engagement and communications plan should be regularly reviewed and adjusted to take into account any changes to planned delivery that need to be communicated.

The BCMs and PrgM should satisfy themselves that people with responsibility for engaging and communicating with stakeholders are doing so. Updates on progress can be included in the highlight reports.

Stakeholder engagement and communications is also a key part of the programme leadership function. The PrgSB and PrgS should champion the programme in day-to-day communication. They should also be sensitive to potential issues at a strategic level and respond accordingly where necessary with input from the PrgM.

5.4.7 Quality management

The implementation phase of the programme ensures that the agreed method for assessing quality assurance and fitness for purpose as agreed during the definition phase is put into practice. This may include activities ensuring that for each project output or programme deliverable, acceptance or performance criteria are in place and are assessed, monitored and reported.

If part of the agreed quality or assurance checks is to undertake periodic reviews at key points in the programme's implementation, these will need to be planned, organised and undertaken. Any findings and recommended actions would be fed into the overall PDP to improve performance.

The PrgM must ensure that when deciding quality and acceptance criteria for deliverables and project outputs, consideration is given to the practicalities of assessing them so that the assessment data can be tested for 'fitness for purpose'.

5.4.8 Procurement and commercial management

The procurement and commercial management of programmes (and its components) will depend on the nature and type of the programme, its funding arrangements, and whether it is in the public sector or in private sector. In some instances, the procurement process may also include obtaining the services of an external PrgM and PMO.

Regardless of the type and approach, the procurement and commercial management processes at the implementation phase would need to consider the following:

- Contract and relevant documentation is in place
- Contractual arrangements are up-to-date
- Business case is still valid and up-to-date
- Original projected business benefit remains feasible
- Processes and procedures are in place to ensure achievement of outputs, outcomes and benefits

- PDP contains processes embedded to ensure all necessary acceptances and fit-for-purpose testing such as: commissioning, soft landings, etc.
- Business contingency, continuity and/or reversion arrangements are in place
- Risks and issues are being managed
- Adequacy and availability of resources
- The delivery plans are still feasible
- There are management and organisational controls to manage the project through delivery and use
- Contract management arrangements are delivery and operation in place to manage the
- Arrangements are confirmed for handover of the project from the PrgS to the customers/client/end user
- There are agreed plans for training, communication, rollout, production release and support as required
- Governance plans are in place
- Information management plans are in place
- Those involved in procurement and commercial management functions need to produce SMART (Specific, Measurable, Achievable, Realistic and Time bound outputs or outcomes) and have a clear understanding of the contract
- The specification, performance measures and contractual terms should be well defined and clear to all relevant stakeholders
- Effective team-working and relationship management need to be in place, and this is especially important in complex programmes due to their long-term nature

5.4.9 Health and safety management

At the implementation phase, the overall health and safety management plan, as prepared during the earlier phases, will be put into action. Typically, all the projects and activities undertaken during the implementation phase will adhere to the health and safety management plan requirements, particularly the targets and KPI's that may have been set at the programme level. The highlight report for the programme may contain a summary of the health and safety performance of the projects and activities, reporting on the actual progress against the targets and KPI's.

Project-level health and safety management will be a key element to ensure achievement of the overall programme targets. *The Code of Practice for Project Management* (5th ed) contains further guidance on management of health and safety at the project level.

5.4.10 Sustainability/environmental management

The overall programme sustainability and environmental management parameters and targets will be set at the earlier phases. At the implementation phase, the processes and procedures, including overall programme targets and KPI's, must be incorporated in individual projects and activities. Projects which may require planning permission will prepare environmental impact assessments that will also encompass the programme sustainability and environmental management aspirations.

5.4.11 Transition management – projects closure

Transition plans are necessary to successfully translate deliverables and outputs into outcomes that will enable realisation of benefits envisaged within the programme business case. These can be included as an update to the overall PDP. Transition plans set out the tasks needed to ensure that the new capability will be embedded into business as usual operations. They might include:

- Training for the new capability
- Providing guidance on the new process
- Allowing parallel running of old and new processes
- Setting up temporary support
- Soft-landing the new capability
- Setting up monitoring processes so that there is assurance that the new ways of working are being used

It is important to note that inadequate transition management may hamper the successful outcome of the programme even if the project outputs are delivered successfully.

6 Stage E: Benefits Review and Transition

> - Do the programme benefits align with the strategic objectives?
> - Does the combination of projects and activities enable optimum realisation of the benefits?
> - Have the programme benefits been clearly described, with sufficient detail, and individual ownership where appropriate?
> - Have the major business changes necessary to achieve the benefits been identified?
> - Have the arrangements been put in place to ensure benefits realisation?

6.1 Purpose of stage

At the point at which the deliverables of a programme are considered by the programme manager (PrgM) and programme sponsor (PrgS) to be achieved, a review is undertaken by the business change manager (BCM) to assess if the required outputs and final outcome have been realised. This review also considers the success of transition of programme outputs into the final permanent state of the new undertaking. This review will confirm whether a programme can be shut down or if further activity needs to be carried out. It is possible that any further activity can be executed without the need to retain the programme management structure.

6.2 Stage outline

This stage (see Figure 6.1) establishes whether the programme has been successful in delivering the desired outcome or whether further deliverables will be needed in order to realise the overall objectives of the enterprise. During the stage the tasks discussed next need to be addressed.

Benefits review

The PrgS, PrgM, BCM and the BRM review the outcomes to ensure that what the programme has delivered is what was stipulated in the programme brief, business case and programme delivery plan (PDP). An assessment is made by the PrgS and BCM as to whether the outcomes delivered by the programme will achieve the benefits required. Their assessment is referred to the PrgSB for ratification.

Code of Practice for Programme Management in the Built Environment, First Edition. The Chartered Institute of Building.
© 2016 John Wiley & Sons, Ltd. Published 2016 by John Wiley & Sons, Ltd.

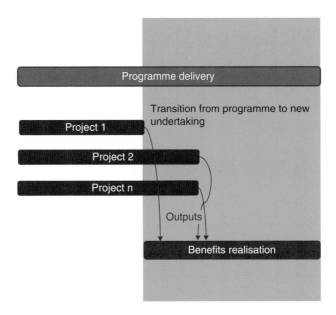

Figure 6.1 Stage E. Benefits review and transition.

There is now an opportunity to make any final adjustments to what has been delivered in the light of greater knowledge and any changing circumstances.

If the PrgSB confirm the programme has delivered what is required, then they instruct the closure of the programme.

Benefits realisation

Realisation of benefits may occur over an extended period as projects become completed, and in some instances the full realisation of the benefits arising from the programme may take months or years beyond the completion of the programme. The situation may therefore arise where it is possible to consider partial completion of a programme and to disband part of the programme team. In these circumstances the BCM will need to maintain an ongoing measurement of the benefits. Also, in some instances this may require the PrgSB to be in place until full and final realisation is achieved.

Transition

Transition from the programme to the new enterprise can be progressive, as elements of the new undertaking are handed over as projects complete, or be a total handover that occurs at the completion of the programme. The new permanent operational state of the undertaking requires a management structure and resources to be mobilised ready to receive and commission it. The transition arrangements are overseen by the BCM.

Training and induction

When the outputs from projects are physical facilities, arrangements must be made for the staff of the new enterprise to receive induction training in the running and maintenance of the new facility. In support of maintaining the new facilities, there should be a transfer of project information, such as design details, building information modelling information, operation and maintenance manuals, legal agreements and design warranties.

Stage E: Benefits Review and Transition

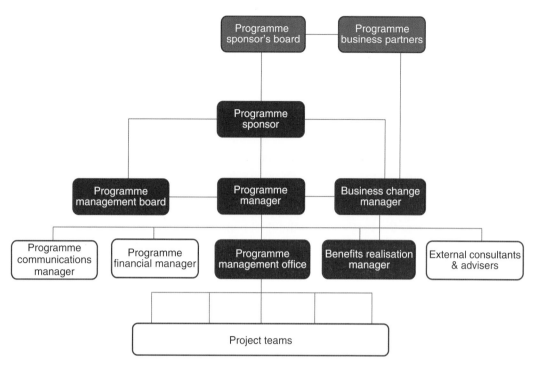

Figure 6.2 Stage E: Benefits review and transition – organisation structure.

6.3 Stage organisation structure

6.3.1 Stage structure and relationships

This stage principally involves the PrgS, BCM, BRM and PrgM (see Figure 6.2) who will establish whether the expected state of the new capability has been achieved. This is done by carrying out the tasks as outline in the next section.

6.3.2 Roles of key participants

Programme sponsor

- In conjunction with the PrgM, initiates the benefit realisation review
- Confirm with the BCM and PrgM that the programme has achieved its objective and can be closed
- Determine if further works are required
- Determine the scope of any further works
- Determine how any additional works will be carried out
- Review with the BCM progress of transition arrangements
- Report to PrgSB on completeness of programme

Business change manager

- Together with the BRM carry out the final benefits realisation review
- Advise PrgS that the programme outcomes which have been delivered will achieve the required benefits
- Advise PrgS of any additional works required to create further benefits

- Manage the transition arrangements for the new capability/enterprise
- Liaise with the client body to ensure their readiness to receive the new capability/enterprise

Benefits realisation manager

- Together with the BCM, carry out the final benefits realisation review
- Determine if there is any benefits shortfall in the programme outcomes delivered

Programme manager

- Assist the PrgS/BCM in the final benefits review
- Assist the PrgS in scoping any additional works required

Programme financial manager

- Advise the PrgS and PrgM on the financial implications of carrying any additional works

Programme management office

- Advise the PrgS and PrgM of the schedule and cost implications of carrying out any additional works

Programme sponsors' board

- Review with PrgS the conclusion of the final benefits realisation review
- Formally confirm the programme has achieved its objectives and can be closed
- Verify any additional works that need to be carried out

6.4 Programme management practices

6.4.1 Benefits management

Benefits management is a critical activity in any programme regardless of its type, objectives and duration. There have been many programmes that delivered great outputs and capabilities but failed to realise benefits due to insufficient or inappropriate arrangements being made to ensure that benefits are realised.

Managing and realising benefits – overview

Benefits management and realisation is a core element of programme (and change) management. It provides a systematic approach to identifying, defining, tracking, realising, optimising, reviewing and communicating benefits, during and beyond a programme.

It is essential that, from the onset, the PrgS takes ownership of the benefits agenda and throughout the life of the programme provides a strong leadership, particularly in terms of prioritising benefits, with support from BCM. Many programmes give in to the temptation of producing a long list of anticipated benefits in the business case in order to maximise the chances of securing funding. Once the funding is achieved, the willingness and commitment to take ownership of all the promised benefits diminishes – the PrgS and BCM should be aware of this and must examine and prioritise the benefits, including assigning individual ownership and seeking commitment from individual benefit owners where appropriate so that benefits realisation is monitored and delivered as anticipated.

The objectives of putting in place processes to manage and realise benefits in a structured way include the following:

- Identification of benefits
- Defining the benefits
- Alignment of the benefits to the strategic objectives, programme vision and outcomes
- Assigning appropriate ownership of benefits, including their realisation
- Aligning project outputs to support the benefits realisation
- Identifying and undertaking business changes that will be necessary to deliver the benefits
- Monitoring and evaluating the benefits and their realisation

By having a structured approach to benefits management, an organisation can ensure that the programme objectives contribute to the organisation's strategic objectives. In addition, it enables the capture of benefits not anticipated at the outset of the programme and provides an ongoing focus, after the programme, on sustaining benefits.

The work to define the programme benefits, develop the PDP and define the project outputs needs to be integrated (see Figure 6.3). Work should start with gaining a better understanding of the benefits outlined in the programme brief and the business changes needed to deliver the benefits. This in turn provides the basis for the scope and the PDP and dictates what the projects need to produce as outputs. The process becomes iterative, as more details about benefits emerge in developing the programme definitions and the project outputs.

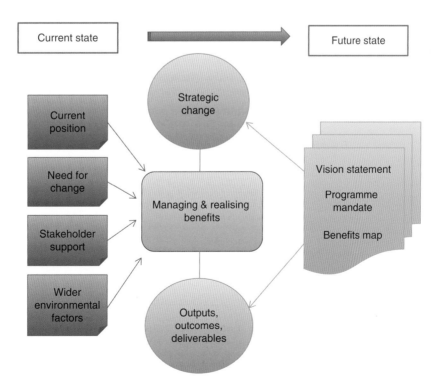

Figure 6.3 Managing and realising benefits.

6.4.2 Benefits and dis-benefits

Benefits

A benefit is a (directly or indirectly) measurable improvement delivered by a programme which is seen by a stakeholder to be positive and worthwhile, for example, creating more green spaces in a particular borough or a perception via a residents' survey that the crime rate has gone down.

Types of benefits

Benefits can be classified in a number of ways. A distinction is made between financial benefits, which are measured in monetary terms, and non-financial benefits, which cannot be measured in monetary terms.

Financial benefits are further categorised as 'revenue', those benefits that give rise to immediate bankable returns, for example, capital receipts from the disposal of property, or 'non-revenue', for example, an efficiency gain leading to less time to complete a required task, which is a gain that cannot be converted into a reduction in staffing or cost.

Non-financial benefits include improvements across services and corporate functions (e.g. human resources, information and communication technologies and legal services) that can be measured using national and local non-financial performance indicators and the results of citizen and staff surveys.

Dis-benefits

Dis-benefits are the outcomes from a programme which are perceived by one or more stakeholders as negative, for example, new operational costs or loss of green space in an area due to the building of a new school. The same change can be seen by different stakeholders as both a benefit (net cost reduction through fewer staff) and a dis-benefit (job losses). These dis-benefits can be classified, managed and measured in the same way as benefits.

Dis-benefits can be confused with risks, but whereas risks may be avoided, dis-benefits will certainly be created by the programme and their impact must be managed. It is important to understand which stakeholder will lose out so that this can be managed. Furthermore, with proactive management, some dis-benefits can be potentially turned into opportunities or, indeed, new benefits (for example creation of a new school may result in loss of green space but extended community use facilities may compensate for the loss).

Benefits identification and mapping

Often, a benefits map is created in a visual form to capture and communicate the benefits. The map can be used throughout the life of the programme to analyse any impacts on benefits caused by changes in programme direction or changes to the strategy as a whole.

There are many ways to map benefits. It is useful to build benefits maps in stages (e.g. start with a session to agree on the programme objectives) because the relationships between the various components can be complex, and it can be better to work on this outside sessions with stakeholders.

The steps outlined below provide an example of how this can be done – the example used in the illustrations considers a leisure facility transformation programme (Figure 6.4a – Figure 6.4e)*.

---

* Source: Public Sector Programme Management Approach – see Bibliography for details.

Stage E: Benefits Review and Transition

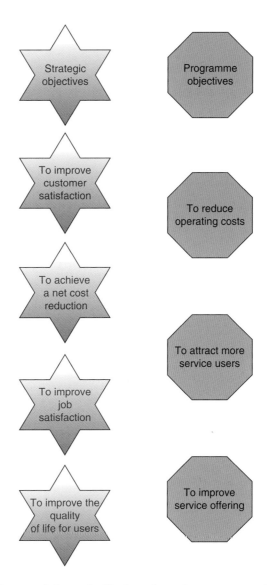

Figure 6.4a Benefits map (leisure facility transformation programme): Step 1 – mapping programme objectives to strategic objectives.

Step 1: Mapping programme objectives to strategic objectives

At the beginning it will be necessary to agree on the programme's objectives if these are not already clear from the programme mandate (see Figure 6.4a). The programme objectives are statements that explain what the programme sets out to deliver at the highest level in the context of strategic objectives.

It is important to ensure that the programme objectives are within the scope and power of the programme.

Step 2: Identifying and mapping benefits to programme objectives

In this step the process involves identifying benefits and dis-benefits. If a programme brief has been created, some benefits and dis-benefits will have already been identified. In addition, more potential benefits may have come to light since the creation of the programme brief.

The identified benefits should then be mapped to the programme objectives (see Figure 6.4b).

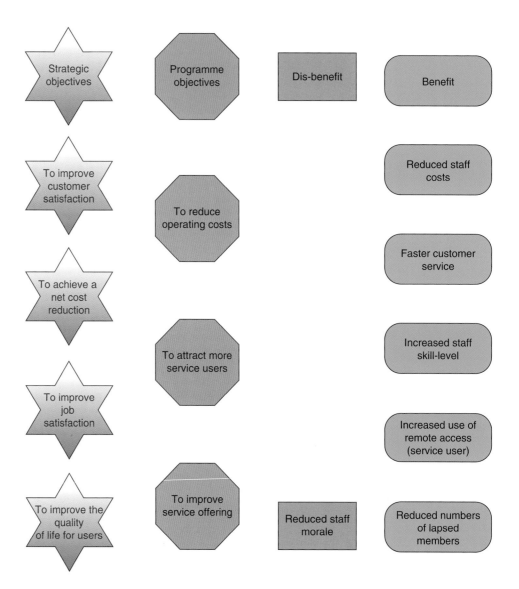

Figure 6.4b Benefits map (leisure facility transformation programme): Step 2 – Identifying and mapping benefits to programme objectives.

Dis-benefits are shown also to the right of programme objectives. It is also useful to indicate if there are any relationships between benefits and/or dis-benefits by placing them adjacent to each other on the benefits map.

Benefits are often intertwined – if relationships between benefits start to become too complex, grouping can be used to identify the related benefits together, and the complexity can be captured in the benefit profiles. In scenarios where there are numerous links between the elements, links can be prioritised in order to maintain visual clarity of the benefits map.

Step 3: Identifying business changes

The next step in the process is the identification of business changes that are needed in order to achieve the benefits (see Figure 6.4c). Business changes are those made to the current ways of working that need to be implemented in the business areas affected by the programme. They can include process and behavioural changes and changes to operational procedures. For example, a new clear-desk policy to support desk sharing will only deliver benefits if staff changes their habits and adequate storage is provided to staff.

Stage E: Benefits Review and Transition

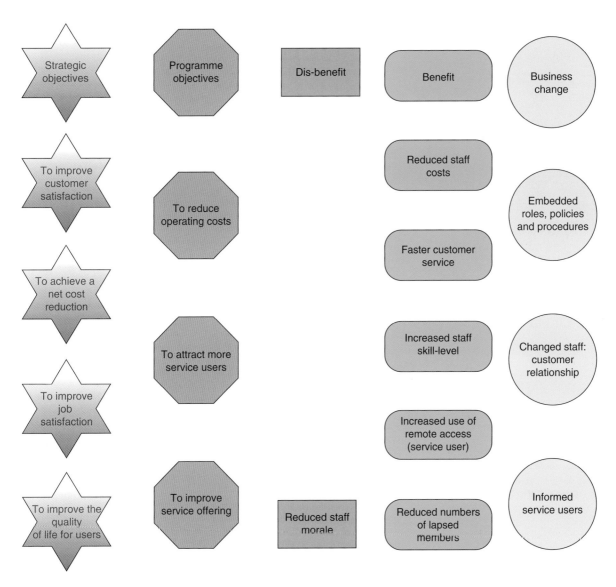

Figure 6.4c Benefits map (leisure facility transformation programme): Step 3 – Identifying business changes.

Typically, the main business changes are noted on the map. A more detailed list of all business changes and how they will be implemented should be developed following the development of the benefit profiles. Business changes are numerous, and it would be difficult to try to fit them all on the map. It is, however, very valuable to use the map to consider all the major business changes required to deliver the benefits.

The activities needed to deliver the business changes should be included in the PDP (or separate benefits realisation plan) or where appropriate within the individual project plans.

Step 4: Mapping project outputs to benefits

The next step is to show the project outputs (enablers) that will create the capability to realise the benefits through the identified business changes (see Figure 6.4d).

In the process of confirming, amending or adding benefits, the project outputs are reviewed to ensure that everything that needs to be created to enable the benefits is listed.

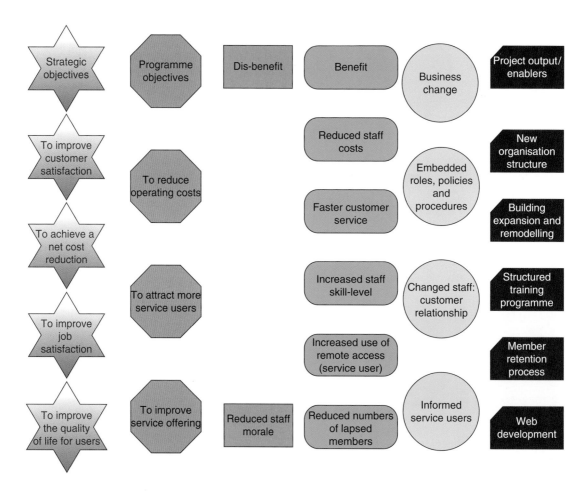

Figure 6.4d Benefits map (leisure facility transformation programme): Step 4 – Mapping project outputs to benefits.

Once the map has been populated with project outputs, business changes, benefits and programme objectives, it can be used as a tool to consider the following:

- Do the programme benefits strongly align with the strategic objectives of the organisation through the programme objectives?
- Will the proposed projects deliver the benefits sought?

It is important to recognise that the benefits map does not indicate when things happen in time, that is, sequence. It is simply a representation of how things are connected to each other.

Step 5: Mapping the links between programme objectives, benefits, business changes and project outputs

Once all elements of the map have been captured, links are identified between the enablers, business changes, benefits/dis-benefits and programme objectives. The number of links can give an indication of the relevant importance of each element on the map and assist with benefit prioritisation (see Figure 6.4e).

Prioritising benefits and business changes

It would not be appropriate or practical to track/measure all benefits. The benefits should be prioritised according to those that are critical to realising the programme objectives, those with the important financial values and also those that are practical to measure (i.e. there is an existing baseline). In practice, this means that only a subset of the benefits are taken forward to the profiling benefits stage.

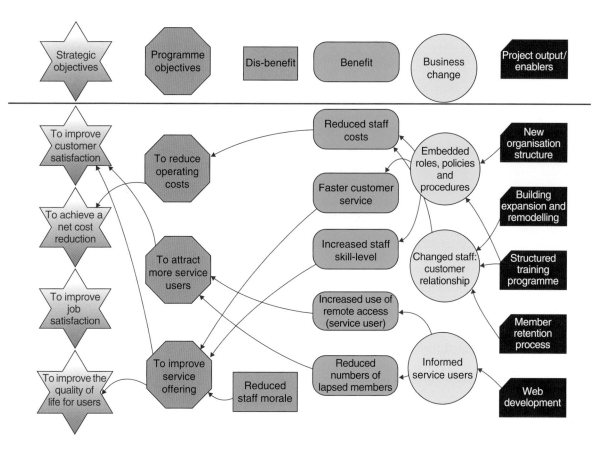

Figure 6.4e Benefits map (leisure facility transformation programme): Step 5 – Mapping the links between programme objectives, benefits, business changes and project outputs.

Profiling benefits

Benefits profiles describe benefits in more detail and record important information. A benefit profile should be prepared for each benefit. This helps to:

- Define the extent of the improvement that the benefit will deliver
- Ensure an appropriate person is accountable for the delivery of the benefit
- Prioritise benefits
- Clarify the project outputs that are needed to enable the benefit to be realised

The detail of benefit profiles will be refined and become clearer as the programme progresses.

See Appendix T7 for a benefits profile template.

Realising benefits

Many programmes fail to deliver their full potential because the responsibility and support is not put in place to deliver all the benefits. The biggest challenge is often to change the way people work.

There are two critical aspects to successful benefits realisation:

- establishing operational benefit owners, that is, people in the service/business who will be accountable for the achievement of the benefits
- implementing the business changes required to embed the new ways of working to deliver the benefits

The BCM role has been developed in programme management to own and lead the realisation of benefits on a day-to-day basis. It is advisable to have an overall BCM who is supported by the benefits owners in the service or business areas involved.

The BCM should plan benefits realisation activities in conjunction with the benefits owners and service areas affected. The following activities should be considered:

- Business change activities, including decommissioning of old ways of doing things, for example, processes, systems, buildings, posts, and implementing new ways of doing things, for example, new processes and training for the same
- Checkpoints for reviewing the changes and resultant benefits
- Responsibilities within the business change area
- Interdependencies and resources required
- Achievement of benefits that are 'early wins' and ways of maintaining focus on benefits that will take longer to achieve

The activities identified should be included in the PDP or included in a separate benefits realisation plan or, where appropriate, they may be included in individual project plans.

Reviewing and embedding

The BCM is responsible for collecting information on the achievements of benefits, and where benefits are not on track, taking the necessary action to ensure they are realised.

There are many ways to do this: See Appendix T8 for an example of a template that can be used to record and track information on all the programme benefits. If required, the financial benefits information can be used for calculating overall financial benefits to date.

The information in a template such as this can also be aggregated with similar templates for other programmes to give a portfolio or strategic view of benefits realisation.

The purpose of tracking is to focus operational business units on achieving, sustaining and improving the benefits. This may mean using existing measures and reports (e.g. from the finance section). It may mean implementing new key performance indicators (e.g. number of lost calls in a call centre operation). In both cases, the measures need to be integrated into the operational management processes and systems, for example, reporting systems, management meetings, personal performance targets, and so on.

By integrating benefits tracking into the existing reporting and management processes, benefits owners have the basis for refining and optimising benefits. This involves looking at the trade-offs between benefits and the impact the operational environment is having on the level of achievement. For example, increasing the number of people working from home by introducing smart-working facilities may reduce office accommodation costs but also reduce effective communication and collaboration between team members.

It is also important to embed and reinforce the new ways of working and discourage the old.

During the life of the programme, benefits should be reviewed by the PrgSB. Structures and mechanisms should also be established, which will continue the process of monitoring and reviewing achievements of benefits beyond the programme closure. The approach, as outlined in pages 100–106, primarily relates to programing a public sector environment (see Public Sector Programme Management Approach – detailed in Bibliography), but can be tailored to suit the specific context of private sector programme as well.

Communications

The communication of the achievement of benefits is an obvious but often overlooked activity. It is important to communicate to stakeholders the success of the programme as it starts to achieve key benefits. Positive feedback and reinforcement is a powerful way to help people transition and adopt new ways of working.

A way of ensuring that the communication of success is not overlooked is to include in the communications and stakeholder plan actions around publicising benefits. Target dates for the achievement of benefits can be taken from the benefit profiles and used to identify when the communications activities should occur.

Sign-off

The key outputs from the benefit management and realisation activities are the benefits map and profiles and the benefits realisation plan (either a separate plan or included as part of the PDP).

The benefits map and profiles and programme plan are signed off by the PrgS during the definition phase.

The benefit profiles and the benefits realisation activities should first be agreed upon by the BCMs (and benefits owners if possible) before they are submitted for sign-off.

The benefit profiles and benefit realisation activities (and plan) need to be kept in alignment with the PDP, project outputs and changes to the objectives of the programme. These documents should be reviewed and changes signed off:

- Before starting a new part of the programme
- When there are major changes to the scope
- When there are developments that impact the attainment of the planned benefits
- Before transitioning to the new state (measure the performance of the current state)
- After transitioning to the new state (measure the performance of the altered state)
- At programme closure to ensure that benefits ownership and realisation continues in business as usual to achieve the benefit targets set out by the programme

6.4.3 Transition strategy and management

Managing the transition strategy should be addressed from (i) an operational, (ii) programme organisation and (iii) business as usual point of view as described below:

I. Operations: Programme close-out/operational readiness

- Assets and facilities ready for operations on time
- All facilities are fit for purpose
- Facilities operate the first time with no critical failures
- Recognised legal, contractual and critical schedule requirements
- Comprehensive operations philosophy and plan in place
- Clear roles and responsibilities pre and post transfer
- Facilities easy to operate and maintain
- Staff to receive timely training to operate and maintain facilities
- Capture of lessons learned for knowledge dissemination

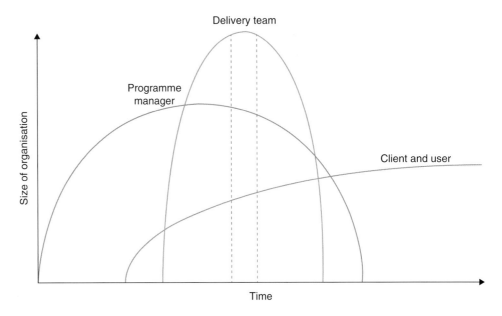

Figure 6.5 Organisation size over time for programme delivery.

II. Organisation: Programme exit strategy

A key challenge is the management of resources to ensure that key staff remains for the life of the wind-up in critical functional areas to ensure that corporate knowledge is retained and there is clear accountability for any associated exit tasks. An interim resource plan using third-party resources could help to shore up unexpected staff losses and provide continuity.

A programme needs to consider how the transition to operations will impact the organisation and the implementation of an exit strategy, as it reaches the end of its life (see Figure 6.5).

Key areas to consider and address:

- Governance: Opportunity to formalise an exit strategy as a formal project/programme
- Resource management: Key staff leaves unexpectedly creating gaps in expertise and loss of corporate knowledge
- Commercial management: Clear visibility of potential contractual issues and tracking of any resulting claims/disputes using a formal system will help to make the exit as efficient as possible
- Legal: Developing a legal risk map will help highlight business requirements for organisational exit
- Stakeholder management and communications: Exit strategy would benefit from the development of a stakeholder map
- Risk management: capturing and formalising will consider all exit strategy risks
- Information technology: Information technology expertise may be required to help support legacy bodies

III. Benefit realisation post close-out

- Handover to BCM to monitor benefit delivery post transition
- Capture of lessons learned for knowledge dissemination

7 Stage F: Closure

> - Will the programme be understood by those looking for lessons learned?
> - Has the programme delivery plan been delivered satisfactorily?
> - Are there any remaining activities? If so, who will do them?
> - Are there any outstanding risks and/or issues? If so, who will manage them?
> - How will the lessons learned be captured and made available?
> - Are there any project activities that will continue? Who will manage them?
> - Does the programme sponsor agree that closure is a true reflection of the status of the programme?

7.1 Purpose of stage

At the point at which the programme sponsor (PrgS) has agreed the required outputs and the outcome defined in the programme delivery plan (PDP) has been achieved and no further works are envisaged, the programme is considered complete.

7.2 Stage outline

The stage involves the controlled shutting down of the programme activities and the disbanding of the programme team (See Figure 7.1).

Shutting down the programme

Prior to disbanding the programme team, a number of final tasks need to be undertaken:

- Lessons-learned review
- Archiving of programme/project information
- Transfer of building information modelling system to management of the new undertaking
- Resolution of any contractual disputes and claims and settlement of final accounts with external parties

---

Code of Practice for Programme Management in the Built Environment, First Edition. The Chartered Institute of Building.
© 2016 John Wiley & Sons, Ltd. Published 2016 by John Wiley & Sons, Ltd.

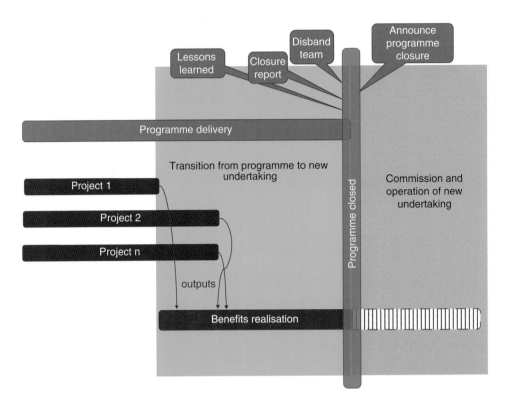

Figure 7.1 Stage F: Closure.

- Final programme communication advising stakeholders of closure
- Preparation of programme closure report and submission to programme sponsor's board (PrgSB)
- Highlighting of any outstanding works or issues to PrgSB

Disbanding of programme team

The manner of disbanding the programme team will vary depending on whether the staffs are permanent employees of the sponsoring organisation or are externally contracted. For those who are permanent employees, a process of co-ordinating the timing of their release and of locating other roles and opportunities needs to be implemented. Contracted staffs leave the programme when the programme manager (PrgM) agrees that this role in the programme has been completed. It is likely the PrgM will stay for a period to liaise with and advise the team taking over the undertaking.

7.3 Stage organisation structure

7.3.1 Stage structure and relationships

Emphasis during this final stage is back with the PrgS who has been instructed by the PrgSB to terminate the programme. The PrgS oversees the final closing down activities and ensures that the business change manager (BCM), on behalf of the programme, is providing appropriate support to the client in the transition period to ensure the successful incorporation of the new capability into the client's business or the formation of a new enterprise. The PrgM, PMO and other members of the programme management team are involved in carrying out various closing down activities.

7.3.2 Stage roles of key participants

Programme sponsor's board

- Instruct the PrgS to close down the programme
- Review closedown reports
- Identify any residual issues that need to be reported to the client

Programme sponsor

- Instruct PrgM to close down the programme
- Ensure all financial matters have been finalised
- Ensure any outstanding claims/disputes have been resolved
- Oversee preparation of a post-implementation review
- Oversee preparation of a lessons-learned log
- Ensure BCM is proving appropriate transition support
- Receive confirmation from PrgM that programme has been shut down and the programme management team has been disbanded
- Ensure all stakeholders have been advised of programme closedown
- Ensure all relevant documentation has been transferred to client/new enterprise
- Confirm with BCM any requirements for future benefits reviews
- Present closedown reports to PrgSB
- Disband the PrgSB
- Ensure any residual issues have been reported to the client
- Ensure any requirements for future benefits reviews have been reported to the client

Business change manager

- Assist in the preparation of the post-implementation report
- Identify lessons learned
- Determine any requirements for future benefits reviews
- Continue to liaise with client regarding the operation of the new capability or enterprise

Programme manager

- Advise the programme team of the termination of the programme
- Ensure all programme and project documentation is complete and archived
- Manage process of termination of involvement/employment of the programme team members
- Confirm process for notifying stakeholders of termination of programme
- Resolve any outstanding claims/disputes with consultant, contractors and suppliers
- Ensure with PrgFM all financial accounts are settled and closed

- Together with programme financial manager, prepare final cost statement
- Co-ordinate the preparation of the lessons-learned log
- Prepare programme post-implementation report
- Identify any outstanding or unresolved issues
- Disband the PrgMB

Programme financial manager

- Together with PrgM, resolve any outstanding claims/disputes with consultant, contractors and suppliers
- Settle and close all financial accounts
- Assist in preparation of final cost statement
- Identify any lessons learned

Stakeholder/communications manager

- In conjunction with PrgM, advise all stakeholders of termination of programme
- Identify any lessons learned

Programme management office

- Ensure the programme and projects information system is complete and archived
- Assist PrgM in preparation of the post-implementation report
- Identify any outstanding or unresolved issues
- Identify any lessons learned

Health and safety manager

- Prepare final report summarising progress of health and safety matters during the programme
- Identify any lessons learned

Sustainability manager

- Prepare final report summarising progress of sustainability and environmental matters during the programme
- Identify any lessons learned

7.4 Programme management practices

7.4.1 Programme closure

The programme closure stage is reached when the PrgS agrees that the required outcomes and outputs as defined in the PDP have been achieved and no further works are envisaged or when there is a significant change on one of the key parameters – for example if there is a strategic change within the organisations that makes the business case no longer viable or the programme funding becomes no longer available.

It is necessary to ensure that a formal programme closure process is undertaken for all programmes regardless of the reason for closure so that there is a formal recognition that the programme has been completed and that delivery of capability and benefits has been assessed. This is particularly important for complex programmes or programmes that have a prolonged delivery period as the organisational 'drift' can otherwise set in, which may allow viewing the programme as part of normal business.

Preparing for closure

Typically, programme closure should happen after completion of the last project or initiative. In deciding whether to close the programme, consideration should be given to whether the business case has been satisfied or is well on its way to being satisfied.

It is often the case, for large and complex programmes, to view programme closure as a separate project where a timeline for closure activities with specific tasks and ownerships are allocated. All relevant parties, including the key stakeholders, should be notified that the programme is being prepared for closure.

A closure activity plan should be prepared to include a programme review and formal sign-off and closure. The following questions should be considered as part of the plan:

- Are there any residual risks or issues that could affect operations?
- Are all the projects complete, including those that are designed to ensure embedding of the new operational environment?
- Has the PDP been achieved?
- Are all other programme activities complete?
- Has the business case been satisfied?

For programmes where closure is necessary due to a fundamental parameter change, confirmation should be sought that all projects have either been closed or, if their outcomes are still required, that alternative and appropriate governing arrangements have been put in place.

Programme review

Before closing the programme, a formal review should take place to assess whether it has delivered what it set out to do. The PrgS should complete the review with input from the PrgM and BCM. The following may be used to test if the programme has delivered against its objectives and outcomes:

- Vision
- Programme brief
- Benefits map
- Business case
- PDP

Some programmes may not have been defined using these specific components. It is most important to review the capabilities that have been delivered and the outcomes and benefits so far.

The following should also be considered:

- Residual risks and issues – have these been assigned?
- Remaining transition activities – have these been allocated?
- Any support functions that require continuation, particularly if relevant to benefits realisation – have these been assigned?

- Any benefits not fully realised – have these been handed over to the relevant business area to be monitored?
- Lessons learned – are processes in place to capture and assimilate?

The programme review must assess the programme's performance and processes to identify and capture lessons learned as these may benefit other programmes.

There may also be a need for independent or external review of the overall programme delivery and performance prior to the formal closure.

Programme closure

Before recommending programme closure to the PrgSB, the PrgS should be satisfied that the business case is delivered with all projects and programme activities complete. In addition, the PrgS should ensure that any required handover or transition activities have been defined and assigned to relevant business functions.

Programme closure should be authorised by the PrgSB on the recommendation of the PrgS. Once the programme's closure activities are completed, the PrgS should then confirm programme closure to all relevant parties. To facilitate the governance requirements, often a closure report (see Appendix T9 for a template) will need to be prepared.

Communications

Communications at programme closure include ensuring that all achievements, that is, delivery of the PDP, outputs, outcomes and benefits, have been advised in an appropriate manner to stakeholders. Where possible, references should be made to the programme initiation documents, including the business case, so that in simple terms it can be set out clearly why the programme was initiated and what have been the final results and benefits.

Disband programme organisation and supporting functions

The programme organisation should be disbanded. This will include releasing all individuals and resources from the programme. Individuals may need to be redeployed back into the organisation, and this should be planned in advance. It is advisable to consider the new skills imparted on the individuals when reassigning them back into the organisation.

Sign-off

The PrgSB authorises programme closure on the recommendation of the PrgS. If the PrgSB are not satisfied with the recommendation, they should give clear direction as to further work required and ensure that resources are available to undertake the work as deemed necessary.

Appendices*

T1 Vision Statement Template

Name of the Programme:
Vision Statement:
Name of the Programme Sponsor:

Version	Date	Status	Author	Change Description

Approved for submission to PrgSB		Date	
PrgSB sign off to proceed		Date	

Distribution List		
Name	**Organisation**	**Role/Function**

Overview

Outline the corporate aspirations, setting out the business intent and benefits being sought.

Business context

Present overview of the state as it is now.

Strategic need

What are the business needs that the organization is aiming to achieve? The needs should be clear and specific, not generic aspirations.

---

* Templates are based on the CIOB code of Practice for Project Management, fifth ed. and Public Sector Programme Management Approach.

Code of Practice for Programme Management in the Built Environment, First Edition. The Chartered Institute of Building.
© 2016 John Wiley & Sons, Ltd. Published 2016 by John Wiley & Sons, Ltd.

Programme vision

How would the future look when the programme is delivered and the benefits achieved? This can be represented quantitatively and or qualitatively.

Constraints and limitations

Include known and foreseeable constraints or exclusions that may apply and any other outcome that will be necessary but not within the programme.

High-level programme scope

Provide an overview of the scope of the programme, outputs and deliverables and list of projects, if possible.

T2 Programme Mandate Template

Name of the Programme:
Programme Mandate:
Name of the Originator:

Version	Date	Status	Author	Change Description

Approved for submission to PrgSB		Date	
PrgSB sign off to proceed		Date	

Distribution List

Name	Organisation	Role/Function

Business need/vision statement

Use this section to set out the drivers that have created the need for this programme. This will include how the programme contributes to the organisation's strategic objectives and fits with other initiatives.

Outcomes

Briefly articulate the outcomes that the programme is expected to achieve. Specify if there are any constraints (e.g. must be achieved by x).

Next steps

List the activities, time and resources required to complete the programme brief and programme delivery plan.

Activity	Time	Resources	Costs

Sign-off

The programme mandate needs to be signed off by the sponsoring board who will commit the resources to develop the programme brief and programme delivery plan.

T3 Programme Brief Template

Name of the Programme:
Programme Brief:
Name of the Originator:

Version	Date	Status	Author	Change Description

Approved for submission to PrgSB		Date	
PrgSB sign off to proceed		Date	

Distribution List

Name	Organisation	Role/Function

Programme vision

Describe a compelling picture of the future that this programme will enable. This should include the new and/or improved services and how they will look and feel and be experienced in the future.

Financial benefits

Describe the measurable improvements that the programme will achieve.

Benefit Description	Current Value	Target Value	Timing	Cashable Value	Non-Cashable Value	Benefit Owner

Non-financial benefits

Describe other benefits that will arise from this programme which are not easily measured.

Dis-benefits

Describe the negative results of undertaking this programme.

Programme activity and projects

Describe the project and programme activities identified so far that will be required to deliver the programme benefits, with estimates of what they will cost and how long it will take to complete the work.

Appendices

Programme Activities				
Programme Activity	**Description/Output**	**Duration**	**Costs**	**Lead Person**
Benefits mapping exercise	Benefits maps Benefits profiles	6 weeks	£2500	A. Smith

Projects					
Project	**Description/Output**	**Duration**	**Contribution to Benefits**	**Costs**	**Lead Person**

Quick wins

State what business activities should start, be done differently or cease, in order to achieve quick wins.

Key risks and issues

List the potential threats (risks) and current issues to the benefits of the programme as they are currently understood. If there is one, use the corporate approach to risk and issues management. This section should be structured according to the corporate guidance.

Risks – Anticipated threats to the benefits						
Description	**Likelihood**	**Impact**	**Proximity (when it is likely to occur)**	**Risk Owner**	**Mitigating Action**	**Action Owner**

Issues – Current threats to the benefits				
Description	**Priority**	**Issue Owner**	**Action**	**Action Owner**

Financial information

- Set out the estimated financial costs and benefits
- List all currently identified or potential sources of funding
- Describe how these figures in the tables below have been arrived at, outlining all your assumptions

Appendices

Financial costs

Estimated Financial Costs – Capital (£000s)					
Description	Year 1	Year 2	Year X	Total	Ongoing
Totals					

Estimated Financial Costs – Revenue (£000s)					
Description	Year 1	Year 2	Year X	Total	Ongoing
Totals					
Total Costs – Capital and Revenue					

Constraints

Describe any known constraints that apply to the programme.

Assumptions

Describe any assumptions made that underpin the justification for the programme.

Programme capability

Describe how the organisation will provide the necessary programme management resources and capability required to carry out the proposed programme successfully.

Sign-off

This section should be signed by a representative of the sponsoring group to confirm acceptance of the brief. Use the version and authority sign-off on the front page.

T4 Business Case Template

Name of the Programme:
Business Case:
Name of the Programme Manager:

Version	Date	Status	Author	Change Description

Approved for submission to PrgSB		Date	
PrgSB sign off to proceed		Date	

Distribution List

Name	Organisation	Role/Function

> This document provides a template for a business case in support of the investment decision.
>
> In some cases, an outline business case may have been completed and agreed upon prior to the submission of this document for approval.
>
> The main purpose of the business case is to provide evidence that the most economically advantageous offer is being pursued and that it is affordable. In addition, the business case explains the fundamentals of the programme and demonstrates that the required outputs and deliverables can be successfully achieved.
>
> For small-scale programmes, a business justification document may be prepared to support the investment decision.[1]

Contents

- Executive summary
- Strategic case
- Economic case
- Commercial case
- Financial case
- Management case

---

[1] See https://www.gov.uk/government/publications/the-green-book-appraisal-and-evaluation-in-central-government for further information and guidance.

Appendices

- Economic appraisals
- Financial appraisals
- Benefits register
- Risk register
- Stakeholder support assessment
- Strategic business plans
- Proposed delivery plan
- Change management plans

Executive summary

An executive summary indicates what decision is being sought and what is the basis of the recommendation.

Strategic case

The strategic case summarises the vision and the strategic drivers for this investment, with particular reference to supporting strategies, programmes and plans.

The strategic context

Contents of this section may include the following:-

- Organisational overview
- Current business strategies
- Other organisational strategies

The case for change

The case for change summarise the business needs for this investment, with particular reference to existing difficulties and the need for service improvement. This should clearly set out the investment objectives and the related benefit aspirations. Contents may include these elements:-

- Existing arrangements
- Business needs
- Potential business scope and key operational requirements
- Investment objectives
- Main benefits criteria
- Main risks with controls proposed
- Constraints and dependencies

Economic case

The economic case should include the options appraised and the outcomes, including the critical success factors. Contents may include the following:-

- Critical success factors
- Long and short list of options
- Economic appraisal of options
- Estimated benefits
- Estimated costs
- Qualitative benefits appraisal
- Qualitative risk appraisal
- Sensitivity appraisal including scenario considerations
- Overall conclusion and recommendation

Commercial case

A commercial case may include the delivery model, risks and contingency options. Appropriate indices may be used to represent non-quantifiable risks and benefits. Contents may include the following:-

- Agreed outputs and deliverables
- Risk management mechanism including risk transfer arrangements
- Contractual arrangements (including risks and personnel issues)
- Delivery timescales
- Accountancy arrangements

Financial case

The financial case should demonstrate the affordability model, where the scheme requires the support and approval of external parties, and indicate that this is committed and forthcoming. A letter of support should be attached as an appendix. Contents of this section may include the following:-

- Impact on the organisation's income and expenditure profile
- Funding and expenditure profile
- Affordability and balance sheet treatment

Management case

The management case should include the management arrangements, including delivery, benefits realisation and risk management. Contents of this section may include these elements:-

- Programme management arrangements
- Project management arrangements
- Use of specialist advisors and consultants

Appendices

- Arrangements for change management
- Arrangements for benefits realisation
- Arrangements for transition management
- Arrangements for risk management
- Arrangements for delivery governance and review
- Contingency plans

T5 Monthly Programme Report Template

Anticipated Final Cost (AFC) by WBS							
WBS	Cost Item (all costs in £m)	Baseline	Earned Value to date	Earned Value for period	Approved Changes	Anticipated Final Cost	Variance
1.1.1	A						
1.1.2	B						
1.1.3	C						
1.1.4	B						
1.1.5	A						
1.1.6	B						
1.1.7	C						
1.1.8	Indirect costs						
	Total						
1.1.9	Contingency						
	Total						

Appendices

T6 Programme Highlight Report Template

Name of the Programme:
Programme Highlight Report No. X:
Name of the Programme Manager:

Version	Date	Status	Author	Change Description

Approved for submission to PrgSB		Date	
PrgSB sign off to proceed		Date	

Distribution List		
Name	**Organisation**	**Role/Function**

Programme Highlight Report		
Report Number	**Date of Report**	**Period (From – To)**

Programme Name:		Ref:		
Programme Vision:				
Programme Sponsor:		Telephone		E-mail
Programme Manager:		Telephone		E-mail

Overall Programme Progress and Status			
	RAG Status		**Comment on overall progress and status and any recommended actions**
	This Period	Last Period	
Time			
Cost			
Delivery/ Outcome/ Output			
Benefits			

1. Overall Programme Financial Overview

Expenditure Type	Total Budget Amount	Total Forecast Spend	Variance Against Budget	%	Spend to Date
Capital					
Revenue					

Comment on financial position and any recommended actions

2. Progress this period

Projects	Detail

3. Milestones Overdue

Project/Sub-Programme	Milestone Description	Expected End Date	Revised End Date	Dependant Tasks/ Milestones? Y/N	Owner

4. Escalated Issues (including those from the last highlight report not yet resolved)

Item	Issue	Recommended Action (s)	Owner
4.1			
4.2			
4.3			

5. Escalated Risks

Item	Risk	H,M,L	Recommended Action (s)	Owner
5.1				
5.2				
5.3				

6. Milestones/Actions for Next Period (in addition to those overdue)

Item	Projects	Activity	Due Date	Owner
6.1				
6.2				
6.3				
6.4				
6.5				
6.6				
6.7				

Appendices

Programme Detail

Performance Against Plans								Financial Performance					Comment
Project/ Sub-Programme	Specific ID	RAG Status						Expenditure Type	Total Budget Amount	Total Forecast Spend	Variance Against Budget	%	
		Time		Cost		Delivery							
		This Period	Last Period	This Period	Last Period	This Period	Last Period						
		G		G		G		Capital					
								Revenue					
		G		G		G		Capital					
								Revenue					
		G		G		G		Capital					
								Revenue					
		G		G		G		Capital					
								Revenue					
		G		G		G		Capital					
								Revenue					
		G		G		G		Capital					
								Revenue					
		G		G		G		Capital					
								Revenue					

Explanation of RAG Status:

Red – Overall slippage
Amber – Slippage in current period
Green – As planned
Grey – Slippage previously reported, recoverable within overall programme
Black – Project completed, on hold or cancelled

Note: The distinction between capital and revenue is only relevant to public sector programmes

T7 Benefits Profile Template

Prepare a benefits profile for each benefit. Benefits profiles describe benefits in more detail and record information to:

- Define the extent of the improvement that the benefit will deliver
- Ensure an appropriate person is accountable for delivery of the benefit
- Prioritise benefits
- Clarify the project outputs that are needed to enable the benefit

Benefit profile:	
Benefit description	*Summary of benefit*
Benefit type	*What are the financial (revenue or non-revenue), non-financial or dis-benefit?*
Programme business changes required	*What are the operational changes needed to achieve the benefits.*
Outputs contributing to this benefit	*What are the activities or projects that, together with the business changes, will create the capability to realise the benefits?*
Benefit owner	*Who will be responsible for making sure that this benefit is realised?*
Stakeholder beneficiary	*Which stakeholders benefit from this improvement (or in the case of a dis-benefit, which will be affected?)*
Measurement and costs	*How will you know that the benefit has been achieved – what measures will you use? This could be existing performance indicators.*
Dependencies on other programmes	*Are other programmes involved in helping to realise this benefit?*
Assumptions	*Are there any assumptions that have been made that underpin the realisation of this benefit?*
Constraints	*Are there any constraints that restrict the level of benefit that can be achieved?*
Risks to benefit	*What are the risks that could prevent this benefit from being realised?*

Measures	Baseline	Target (s)	Measurement method and responsibility
Description of measure	*Starting point from which you will measure this benefit*	*Target value and timescale (there could be several interim targets until you achieve the final target)*	*How the benefit information will be captured and who is responsible*

Benefit profile:			

Sign-off

The benefits profiles should be signed-off by the PrgS (or BCM on their behalf) to confirm acceptance of the benefit profiles.

T8 Tracking Benefits: Benefits-Monitoring Template

	Benefit ID/name	Owner	Baseline	Target value	Target date	Last review and achievement to date	Next review	Action points with ownership
Financial revenue								
Financial Non-revenue								
Non-financial								

Appendices

T9 Programme Closure Report Template

Name of the Programme:
Programme Closure Report:
Name of the Programme Manager:

Version	Date	Status	Author	Change Description

Approved for submission to PrgSB		Date	
PrgSB sign off to proceed		Date	

Distribution List

Name	Organisation	Role/Function

Context

Use this section to outline the history of the programme, a few paragraphs that include why it was needed, when it started and the reason for its closure.

Delivery

Outline how much of the programme has been implemented. If not all of its components have been created, outline the reasons for the shortfall; if the partially changed state of the organisation requires any further activities, where does the responsibility for those activities now lie.

Benefits

Using the table below, for each benefit list the benefit measures that have been captured to show how well the business case has been achieved. The first three columns are copied from the benefit profile.

Benefit Description	Target Value	Achieved value	Target Date	Achieved Date

Handover

For any benefit that has yet to be fully realised, list who is the new owner of the benefit realisation activities and whether this responsibility has been formally handed over and accepted.

Risks

List all outstanding risks and where the ownership now lies.

Risk	Previous Owner	New Owner

Issues

List all outstanding issues and where the ownership now lies.

Issue	Previous Owner	New Owner

Projects (optional section for premature closure only)

Normally, all projects will be closed at the end of the programme. In the case of premature closure, there may be some projects that will still be valuable. List the existing projects that will still be useful to the organisation and where the new ownership now lies.

Project name	Reason for Continuing	New Owner

Lessons learned

Highlight key lessons learned (positive and negative) that should be passed to on-going and future programmes. Consider the following:

- Governance organisation
- Stakeholder engagement and communications
- Vision and blueprint creation and delivery
- Benefits realisation
- Business case management
- Financial management
- Resource management
- Programme risk and issue management
- Programme planning, monitoring and control
- Quality management
- Change control

Sign-off

The closure review needs to be signed off by the PrgS, who will report to the PrgSB in order to gain approval for the formal closure of the programme.

Use the version and authority sign-off on the front page.

Key Roles: Skills and Competencies

Programme manager

Main duties

- Lead and direct the programme management team comprising:
    - Planning and control
    - Cost
    - Finance
    - Risk and opportunity

- Other programme support as required from time to time such that the programme team successfully provide the required support, guidance, analysis and advice at project and programme level as required

- Building and maintaining strong relationships within the client organisation to ensure that the programme team integrates with the client organisation, addressing any cultural and procedural issues that may arise with integration in a positive and constructive manner

- Implementation of a structured programme management methodology including supporting processes, procedures and tools

- Analysis of information from projects at programme level, with outcome reflected in period reporting to the programme sponsor. Analysis to reflect sensitivity of information and to include recommendations for actions, covering cost, funding, risk and opportunity, time, quality, interdependency and so on

- Continually seek to identify and fulfil client requirements and meet them in innovative and structured ways that add value and increase the probability of success of the programme

- Positive promotion of the programme to all key stakeholders

Key competencies

Leading others

- Works across boundaries sharing information and matching resources to priorities

- Is a visible leader who inspires trust, actively uses teamwork to deliver objectives and takes responsibility for overcoming setbacks

- Communicates at all levels and with diverse groups and able to present complex information clearly

- Is honest and realistic providing clear direction, focusing on strategic outcomes.

Managing people and performance

- Communicate and agree on measurable objectives with teams and staff

- Manage change and continuous improvement dealing with resistance and conflict in a constructive way

Project and programme management

- Make cross-cutting connections between issues and departments
- Use communication strategies to present ideas in a clear and positive way
- Be aware of the wider political environment
- Analyse and use evidence
- Use evidence to evaluate projects and programmes
- Engage with relevant specialists to supply and evaluate all evidence

Financial management

- Ensure that agreed objectives are delivered on time and within budget
- Interpret trends and risks in financial management reports
- Understand the wider expenditure and financial decision-making environment
- Set targets to improve value achieved from resources

Key criteria

Key communicator recognises that the role of programme manager demands regular contact and negotiation with the team, clients, consultants, contractors and other stakeholders. Focus on customer satisfaction is paramount.

Essential skill is to be able to see the big picture; recognise the detail of projects and how they could contribute the success or otherwise of the programme.

Business skills include cost funding reconciliation reporting, HR management and leadership skills, budget information and control, including forecasting, payment processes and so on.

Strategy and planning

Project and programme management skills

- Implementation and monitoring adherence to programme management procedures
- Sound technical knowledge in a variety of disciplines involved in the delivery of the projects and programme
- Understanding of what is involved to identify and develop potential projects from feasibility through evaluation, culminating in the production of a business case, including investment appraisal and identification and implementation of most suitable procurement strategy for a portfolio of projects
- Proven, successful experience in the management of alliance or partnering working, co-locating a team within a client organisation
- Successfully coordinating activities including the ability to organise the workload of the team, balancing priorities and scheduling resources; able to deal with problems on own initiative and to make sound and timely decisions on a day-to-day basis
- Ensuring the co-ordination and identification of programme risks and management of risks through development and maintenance of relevant tools

Business change manager

Main duties

The business change manager (BCM) role is mainly benefits focused. The BCM is responsible, on behalf of the programme sponsor, for defining the benefits, assessing progress towards realisation and achieving measured improvements. The BCM role is associated mainly with programmes that tend to be more benefits focused than projects, although projects that deliver benefits in their own right may warrant the creation of a BCM role.

The BCM role must be the 'business side' in order to bridge the programme and business operations. Where the programme affects a wide range of business operations, more than one BCM may be appointed, each with a specific area of the business to focus on.

The BCM is responsible for the following:

- Ensuring that the interests of the programme sponsor(s) are met by the programme
- Obtaining assurance for the sponsoring group/programme sponsor that the delivery of the new capability is compatible with the realisation of benefits
- Working with the programme manager to ensure that the work of the programme, including the scope of each project, covers the necessary aspects required to deliver the products or services that will lead to operational benefits
- Working with the programme manager to identify projects that will contribute to realising benefits and achieving outcomes
- Identifying, defining and tracking the benefits and outcomes required of the programme
- Ensuring that maximum improvements are made in the existing and new business operations as groups of projects deliver their products into operational use
- Leading the activities associated with benefits realisation and ensuring that continued accrual of benefits can be achieved and measured after the programme has been completed
- Establishing and implementing the mechanisms by which benefits can be delivered and measured
- Taking the lead on transition management, ensuring that business as usual is maintained during the transition and the changes are effectively integrated into the business
- Preparing the affected business areas for the transition to new ways of working
- Optimising the timing of the release of deliverables into the business operations

Key competencies

The individual appointed as BCM should be drawn from the relevant business areas wherever practicable. Their participation in the programme should be an integral part of their normal responsibilities to enable changes resulting from the programme to be firmly embedded in the organisation.

BCMs require detailed knowledge of the business environment and direct business experience. In particular, they need an understanding of the management structures, politics and culture of the organisation owning the programme. They need effective marketing and communication skills to sell the programme vision to staff at all levels of the business, and BCMs ideally should have some knowledge of relevant management and business change techniques such as business process modelling and re-engineering.

Benefits realisation manager

Main duties

The benefits realisation manager role is responsible for identifying, base-lining, profiling, planning, tracking and reporting the benefits. The role involves developing and then managing the processes and management systems needed to support and govern effective benefits enablement and realisation to ensure the programme meets its objectives and realises its target financial savings

The role is responsible for embedding and aligning the concept and principles of benefits realisation and contributes to a change in culture and behaviour across the programme in respect of benefits management and trains, educates and mentors, where appropriate, those staff directly involved in the delivery of business benefits.

- Develops and supports the benefits management strategy and ensure that it reflects the direction of travel within the business and continues to be fit for purpose
- Defines the benefit policies and procedures for the organisation
- Defines, evaluates, recommends, monitors and assures benefits derived from component projects and the overarching change portfolio across the whole investment life cycle
- Defines, manages and updates the organisation's benefit map against investment outcomes, profiles, interdependencies and realisation plans
- Provides assurance that all selected component projects are aligned to the agreed benefits strategy and map and any impact identified
- Provides the cost-benefit analysis data of the component projects' business cases and how these align to the portfolio benefits map
- Supports management's decision-making by analysing benefit options and predicting future costs
- Supports strategic business change by developing working practices that link benefits management into efficiency planning, performance measurement and 'value for money' delivery, ensuring benefit-led project prioritisation
- Monitors benefit realisation plans and benefit review schedules
- Ensures benefit owners are in place and the benefits are profiled, communicated, understood and being managed
- Analyses variances and initiates corrective actions with the benefit owners
- Reviews the impact on the organisational benefits of new projects and change requests
- Provides assurance to the organisation that the benefits are measurable, realistic and achievable and that the risks to the benefits are being effectively managed
- Expedites delivery of benefits by establishing and maintaining working relationships with sponsoring board, business change managers, project/programme managers and other key stakeholders to ensure the benefits are planned and realised
- Initiates benefit reviews to provide assurance of benefit realisation plans
- Is responsible for enhancing the organisational understanding and knowledge of benefits management

- Maintains industry standard professional and technical knowledge
- Prepares reports by collecting, analysing, and summarising information and trends as requested by the programme management office or other performance/governance bodies
- Monitors benefit trends and analysis methods from other organisations
- Attends relevant project and programme boards and departmental meetings to provide updates on benefits management and to provide practical advice to support delivery
- Identifies proactively business benefit opportunities by liaising with key organisational stakeholders and assists, offers advice and guidance to enable a systematic business benefits process to be established
- Monitors benefit realisation activity after component project delivery
- Reviews benefits realisation achievements and puts continuous improvement processes in place
- Identifies benefits within the various stages of the business case development

Key competencies

- Have training in benefits realisation management with 5 years plus experience in a relevant field
- Have recent financial accountancy experience
- Demonstrate experience with the development of benefits management strategies, techniques, processes and tools
- Demonstrate experience of cost-benefit analysis methods, benefit mapping and benefit-profiling tools
- Have proven record of stakeholder engagement and working directly with executive teams, programme sponsors and corporate finance
- Have recent experience of the development and implementation of management information processes and products related to benefits realisation
- Able to apply structured business improvement techniques to identify business benefits
- Have financial accountancy experience
- Able to understand the strategic aims and objectives of the organisation
- Demonstrate strong numerical and verbal critical reasoning ability
- Demonstrate strong financial accountancy skills in terms of defining and projecting future benefits and associated costs
- Able to analyse both qualitative and quantitative benefits information
- Possess a high degree of accuracy and attention to detail
- Demonstrate leadership of, and a positive approach to benefits management, demonstrating a willingness to challenge existing practises to support the organisation to continuously deliver benefits
- Able to mentor and coach project managers and other practitioners in the benefits management processes

- Demonstrate experience and competence in the use of MS Office applications (specifically Word, Excel and PowerPoint)
- Demonstrate a personal commitment to own professional development
- Able to recognise where processes are required and to develop and improve existing processes

Programme financial manager

Main duties

- The core responsibility of this role is the co-ordination, control and reporting of cost information related to the programme
- Client reporting requirements include working with project teams and finance to review delivery organisation information and then assemble individual project reports, including quality check; working with programme manager to compile overall programme report and quality check; ensure reports meet customer requirements, including analysis of costs at project and programme level, variance analysis, forecasts, cost plans, budgets requirements, third-party funding and so on and implement agreed performance indicators to monitor projects more effectively; in addition, ensure substantiation and audit trail is maintained
- Supervise other commercial resources that may be supplied by the programme office to ensure reporting requirements for all projects within the programme are met
- Interface with head of programme, programme managers and project management teams and manage cross-project dependencies from a high-level business perspective
- Implement structured programme management cost information gathering and reporting methodology, including supporting processes and procedures and tools
- Promote the programme to all key stakeholders
- Build and maintain strong relationships with senior colleagues within the client organisation
- Support project management teams to ensure successful project delivery
- Contribute with other programme managers for planning and control, cost and finance to ensure accuracy and uniformity of project reporting across the programme
- Assist in supporting interface with finance for affordability analysis across the programme and other corporate information exercises
- Contribute to the programme risk and opportunity review by identifying possible conflicts and synergies visible through commercial analysis
- Implement a structured programme management cost information gathering and reporting methodology, including supporting processes and procedures and tools

Key competencies

- Significant commercial and cost management experience with proven track record of cost planning, monitoring and control, applying tools, principles, skills and practices to major programme of work

- Knowledge of programme and project management methodologies and experience in tailoring generic approaches to practical business situations
- Effective interpersonal and communication skills, verbal as well as written
- Able to find ways of solving or pre-empting problems and flexibility to be able to react to change in a positive manner

Key criteria

- Estimating and/or reviewing capital cost estimates
- Identifying and/or reviewing of operating cost
- Evaluating different funding sources and making recommendations
- Managing cash flow (aligned with project schedule)
- Projecting revenue
- Assessing risks/opportunities and associated cost
- Evaluating procurement strategies and making recommendations
- Identifying programme level savings (supply chain, strategic purchases)
- Assessing indexation/inflation
- Managing and controlling cost including change control
- Implementing common CBS/WBS/OBS required for programme
- Assessing trends, sampling, measuring, benchmarking, whole life costing
- Providing business case analysis, project gateway reviews, lessons learned

Head of programme management office

Main duties

- Monitoring, independently reviewing, and reporting on the delivery of the programme
- Establishing robust programme delivery reporting across the domain using the existing system, reports and tools available or to set up new systems where necessary
- Establishing independent health check criteria on programmes that will provide an independent view of delivery successes, risks and issues
- Performing regular independent health check reviews on a material portion of the programme
- Setting up and running high-level independent health check meetings
- Updating the programme sponsor on overall programme delivery, identifying key delivery challenges, and proposing viable solutions to risks and issues in conjunction with programme managers
- Ensuring programme static data and ongoing delivery updates are accurately captured in clarity, useable for the appropriate audience, and beneficial for future planning and prioritisation exercises
- Managing the team of PMO resources
- To act as the trusted partner and adviser around programme delivery

Key competencies

- Have a technology background, including both programme/project management and application development experience
- Have management experience working within a change division
- Demonstrate an understanding of programme management and change practices
- Show a desire to provide independent, agnostic oversight on a large portfolio of programmes separate from the teams actually owning and delivery of the programmes
- Have technical development experience, an essential but not primary focus of the role

Key criteria

- Educated to degree level in technology or engineering from a university
- Experienced in a financial services institution desirable but not mandatory
- Able and willing to manage and control detailed metrics, risks, and issues related to technology programme delivery
- Experienced in using management information software application is desirable
- Experienced with corporate strategies and organisational structure, and business policies and procedures in order to provide senior leadership

Risk manager

Main duties

- Produce and manage the risk register
- Own and promote the risk management process as defined in the quality management system
- Produce and manage the risk register
- Advise project teams on 'best practise' project risk and opportunity methodology
- Conduct quantitative schedule risk analysis on the schedules
- Conduct quantitative cost risk analysis on the cost plans
- Assist project teams to manage project risks as part of the individual project risk registers
- Attend and participate in risk and opportunity management workshops where required by the project teams

Key competencies

Leading others

- To promote risk and opportunity management throughout the programme and projects teams
- To act as point of contact for all technical/specific risk and opportunity management related queries

- To produce programme risk and opportunity management process proposals and communicate

- To coordinate management of programme risk and opportunities register

Project and programme management

- To ensure risk management methodology is incorporated fully within programme and project management

- To offer advice and recommendations on risk plus opportunity management to project teams

- To understand how risk management fits into the overall project and programme lifecycle

Analysis and use of evidence

- To understand risk and opportunity products and processes to inform project and programme reports

- To understand programme implications of projects risks

Key criteria

- Project risk management skills

- Technical risk skills such as use of risk management software, risk modelling skills, risk analysis skills

- Understanding of project management lifecycle, planning and cost principles

- Workshop facilitation skills

- Presentation skills

- Report-writing skills

- Problem-solving skills

- Communication

- Process implementation management

Scheduling manager

Main duties

- To set up, maintain and use the master programme schedule and key milestones from the individual projects analysed at programme level to identify cross-project critical path, schedule risks, resource peaks/troughs and so on. This will involve detailed and regular communication with the project management teams to ensure that robust bottom-up information is flowing from the projects to enable programme level analysis

- To analyse potential problem areas – observations and recommendations for action will be required. Establish cost on cash flow and funding, commercial risks, cross-reference with risk for schedule risk analysis and so on

- To manage schedule control system (master schedule), conceived to serve as the management tool for planning, monitoring and controlling the design, procurement and construction of the individual projects and overall programme at a strategic level

- To achieve the programme goals through the development of a well-defined and realistic plan

- To provide a visual means of conveying the information contained in the plan to stakeholders

- To facilitate regular updating and monitoring of the programme

- To prepare a master schedule that contains all relevant time schedule information from the individual projects

- To ensure that the project teams update their current schedule with actual executed information and submit this each period. Review the submitted reports and compare with the master schedule. Assessments sheets will be produced four weekly

- To examine the period progress report from the delivery organisation project teams by the responsible project manager in conjunction with the relevant key personnel

- To update the master schedule on a monthly basis. The master schedule will form part of the periodic report to the programme board

- To undertake milestones trend analysis to identify all relevant project and programme milestones

- To prepare a cost-loaded programme in an agreed level to get qualified and schedule-interdependent information for the cash flow

- To connect the information for risk to the master programme

- To add to the general section of the master schedule to reflect key decision points and milestones

- To continue communication with project managers to remind them of standard and frequency of time schedule reporting

Key competencies

- General understanding of the interfaces and interdependencies between the projects/departments

- Qualities to lead people

- Highly effective interpersonal and communication skills

- Ability to find ways of solving or pre-empting problems

- Flexibility to be able to react to change in a positive manner; willingness to provide support in areas outside core role for the overall benefit of the programme management team

Key criteria

- Significant schedule management experience with proven track record of activity planning, monitoring and control applying tools, principles, skills and practices to major programme of work

- Knowledge and experience in the established tools from Microsoft and Primavera software
- Knowledge of programme and project management methodologies and experience of tailoring generic approaches to practical business situations

Cost manager

Main duties

- Check/validate applications for payment from delivery organisations with reference to numerical accuracy, allowable/disallowable costs, duplication, valuation against progress, valuation against elements of project
- Advise the project manager and assist with resolution of anomalies
- Assist finance with checking/validation of Invoicing from delivery organisations
- Review and confirm (as deemed appropriate) delivery organisations' estimated costs, operational expenditure, revenues and forecasts
- Work with delivery organisations to increase levels of confidence in financial information being provided both on a periodic, annual and out-turn basis
- Act as 'bridge' to facilitate better reporting of commercial issues and their subsequent financial impact
- Review (and align) delivery organisations' accruals methods
- Facilitate early intervention through speedy identification of issues affecting projects with major implications for project financing
- Identify early any issues relevant to current funding/budget availability and help provide clarity on impact of issues such as rollover or transfers on budget
- Deliver project cost information that is accurate, timely and reliable
- Ensure tight commercial, financial and business controls are in place
- Measure project performance against objectives, forecasts and budgets
- Assist in the clear presentation of project expenditure
- Prepare and analyse periodic cost reports at the programme level
- Identify process improvement opportunities

Key competencies

- Understand applications for payment and invoice process – to liaise with delivery organisations and project teams to ensure TS pay only for valid services or products and to ensure value for money
- Have experience in management accounting, including project reporting – to review and interrogate delivery organisations' reports to TS to ensure accuracy, consistency and completeness
- Understand capital grant funding, RAB financing and other available funding options and the associated reporting requirements – to enable preparation of the TS affordability models
- Interface between financial and project/programme teams

Key criteria

- Have experience preparing project accounting information
- Have experience with contracting or project environment
- Be commercially aware, with strong analytical and communication skills
- Be adaptable and a motivated self-starter
- Have strong interpersonal skills with non-financial management
- Demonstrate planning, organisational and analytical skills
- Have high degree of computer literacy, including spreadsheet and MS office skills

The skills, competence and key criteria for the roles outline are for the guidance purposes only. The specific will vary depending on the individual requirements and context of each role and programme.

Programme Management Case Studies

> *Case Study 1 – Example of a Vision-Led Programme: London Olympics 2012*
> *Case Study 2 – Example of an Emergent Programme: High Street Retail Store Re-branding*
> *Case Study 3 – Example of an Emergent Programme: Highways England*

Case Study 1 – Example of a Vision-Led Programme: London Olympics 2012

1 Introduction (organisation and programme)

London's successful bid for the 2012 Olympic and Paralympic Games created the need for a major regeneration and construction programme to provide the venues and infrastructure needed to stage the Olympic Games.[1] The programme of construction was extensive and technically and politically challenging. It was also up against a fixed deadline of the Opening Ceremony of the Games in July 2012. Turning the vision of the Olympic bid into the reality of roads, bridges and stadia was the job of the Olympic Delivery Authority (ODA), a new, publicly funded body established by an Act of Parliament in April 2006.

2 Programme description

2.1 Aim/objectives

Vision statement: "We will use the power of the Games to inspire change."

Programme goal: "To host an inspirational, safe and inclusive Olympic Games and Paralympic Games and leave a sustainable legacy for London and the UK."

Strategic objectives

- To stage an inspirational Olympic Games and Paralympic Games for the athletes, the Olympic and Paralympic family and the viewing public;

- To deliver the Olympic Park and all venues on time, within agreed budget and to specification, minimising the call on public funds and providing for a sustainable legacy;

- To maximise the economic, social, health and environmental benefits of the Games for the UK, particularly through regeneration and sustainable development in East London; and

- To achieve a sustained improvement in UK sport before, during and after the Games, in both elite performance – particularly in Olympic and Paralympic sports – and grassroots participation

---

[1] Case study: Programme Management available at http://learninglegacy.independent.gov.uk/publications/programme-management.php.

Appendices

2.2 Content, scope and scale

The programme scope, broadly, was to procure and deliver £6 billion of major construction works, comprising:

- The deconstruction and land remediation of an approximately 400 hectare site
- Planning and submission of approximately 950 individual planning applications
- The delivery of the infrastructure (20 km roads, 13 km tunnels, 26 bridges, new utilities ([gas, water, electricity, drainage and telecommunications])
- Delivery of 14 permanent and temporary sporting venues; 12,000 sqm broadcast centre and 29,000 sqm media centre (for commercial use as a legacy benefit)
- Delivery of the athlete's village (converted to 2,800 homes as a legacy benefit)
- Creation of approximately 100 hectare of parklands, gardens and public open space
- Transport improvements including station and infrastructure works

The ODA budget was set at £8 bn in 2007 including a £2bn of contingency, with the overall delivery budget of approximately £9.298 bn.

2.3 Duration/time line

The timeline for the programme and the individual phases is summarised below:-

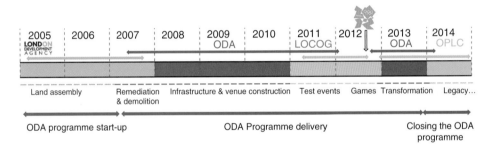

Organisation	Period	Activity
London Development Agency (LDA)/ODA	Year 1 (2006–2007)	Planning and land assembly
Olympic Delivery Authority	Year 2 (to Beijing 2008)	Demolish, dig, design
	Year 3 (to 27 July 2009)	'The big build' (foundations)
	Year 4 (to 27 July 2010)	'The big build' (structures)
	Year 5 (to 27 July 2011)	'The big build' (completion)
London Organising Committee of the Olympic and Paralympic Games	Year 6 (to 2012)	Testing and commissioning, the Olympic and Paralympic games
Olympic Park Legacy Company	2012 onwards	Conversion of the Olympic park and venues to permanent legacy configuration, re-opening the Park to the public

3 Programme organisation

3.1 Overall structure and relationships

Although delivery responsibility was cascaded down to project level, all key policies and processes were determined and implemented at a programme level. All projects followed a standard approach for governance, control and reporting mechanisms.

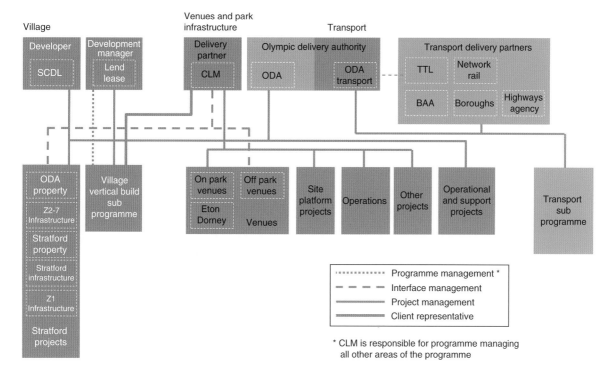

3.2 Roles of key participants and responsibilities

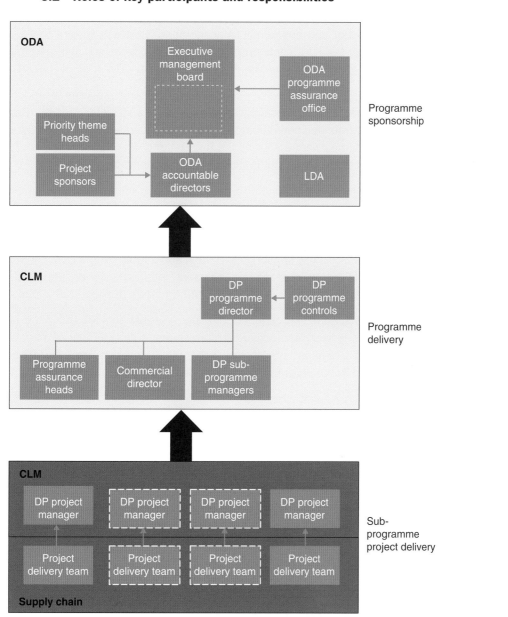

3.3 Key stakeholders

Responsibility for preparing and delivering the Olympic and Paralympic Games and their Legacy resides with the lead delivery stakeholders, namely LOCOG (London Organising Committee of the Olympic Games and Paralympic Games Ltd.), ODA (Olympic Delivery Authority), GLA (Greater London Authority), LDA (London Development Agency), BOA (British Olympic Association), BPA (British Paralympic Association), OPLC (Olympic Park Legacy Company), OSD (Olympic Safety and Security Directorate) and HMG (Her Majesty's Government - GOE [Government Olympic Executive]).

4 Programme management process and practices

4.1 Key stages and management practices conducted in each stage

From the outset, ODA instituted processes and systems, meeting structures and delegations to ensure strategic direction, performance management and value for money. The delivery partner (a private sector consortium CLM comprising a partnership from the three parent companies of CH2M, Laing O'Rourke and Mace) was fully integrated into all aspects of governance and was appointed to be the ODA's overall programme management partner, as well as have the responsibility of project manager for the major construction projects: this was preferred over the alternative approach of separating project and programme management, as it was judged that there were clear benefits of having a common programme and project manager and that the potential conflicts of interests could be managed by having ODA in an overall assurance role.

Delivery Partner (DP) was responsible for leading programme and project delivery review meetings; stakeholder meetings were also held on every project. For each priority theme, dedicated boards were formed to assure that at a programme level the priority themes were achieving the targets, and where shortfalls were forecast, appropriate measures were put in place to enable delivery.

There were several aspects to managing integration across the programme – these included dependency management, design management, physical integration and change management. Across all aspects of managing integration, the two key elements were issues identification and issues resolution; at project level these were managed within the project governance, and at programme level additional processes were required to manage cross-project interfaces to ensure that consequential impacts were identified and programme priorities applied in resolution.

Where escalation was required, this was assessed in context of the overall impact – both from delivery and consequential issues, a key element of managing this having been the integrated programme schedule. Decisions were prioritised and options assessed in the programme (as opposed to an individual project) context.

5 Benefits realisation

The majority of benefits generated by the ODA Programme delivering venues and infrastructure are to be realised after the transformation from Games format to legacy, through future sporting and regeneration uses. The management of the direct benefits obtained from hosting the London 2012 Games was a responsibility of LOCOG and HM Government (DCMS-GOE). The legacy benefits are to be delivered by other agencies, such as the Lea Valley Regional Park Authority (LVRPA) and the Olympic Park Legacy Company (OPLC).

The project business cases contain a summary of the anticipated benefits for each of the projects. At the close-out stage, ODA reported on any benefits achieved and identified owners from other stakeholder organisations for the realisation of future

Appendices

benefits. The process for transferring the benefits was managed by the GOE through the Evaluation Steering Group.

In addition to the benefits identified in individual business cases, London 2012 also appeared to have delivered a number of unintended benefits. These include benefits obtained through coordination exercises (particularly in context of organisations who normally would not work together), communication benefits (in terms of a variety of partnerships working together) and enhanced delivery capacity for participating organisations.

6 Lessons learned

A report by the Department for Culture, Media and Sport[2] identified that the Games have enabled a number of lessons to be learned about how to maximise the benefits to the host country and city from the staging of mega-events. These include:

- Setting a clear legacy vision at the outset
- Political commitment
- Clear remit and accountability structures
- Engagement and participation of key stakeholders from the outset
- Adequate funding to support the objectives
- Ability to remain flexible and change plans
- Sustaining momentum and focus once the projects are finished
- Shared agenda and common understanding across the delivery organisation and key stakeholders

---

[2] *Report 5: Post-Games Evaluation - Meta-Evaluation of the Impacts and Legacy of the London 2012 Olympic Games and Paralympic Games Summary Report* available at https://www.gov.uk/government/uploads/system/uploads/attachment_data/file/224181/1188-B_Meta_Evaluation.pdf.

Appendices

Case Study 2 – Example of an Emergent Programme: High Street Retail Store Re-branding

1 Introduction

Following the introduction of a new-generation design of display unitary and finishes for the retail floor space of a franchised high street retailer, a board level directive required that 100 stores of the group's 550 store portfolio were to be re-branded by the end of that financial year – a period of just 5 months. The retail organisation had an existing, experienced in-house project management team structure but it was already committed to delivering existing projects. With relative low technological complexity, the challenge of this undertaking was the severely limited time for delivery, the number of projects to be executed and the organisational complexity created by the number of interested stakeholders. The successful solution was the adoption of programme management to provide centralised strategic direction and control mechanisms.

2 Programme description

2.1 Aims and objectives

Vision statement – To increase sales turnover by enhancing the attractiveness of the customer store experience.

Programme goal: To effectively re-brand 100 stores within a 5 month period.

Strategic objectives

- Allow the 100 store target to be completed in the given time
- Permit the projects to be 'stand alone' and not be part of the existing re-brand programme
- Ensure a seamless project delivery to every store involved – these were based in England, Scotland, Wales, Northern Ireland and the ROI
- Allow all necessary approvals to be given with minimum or no delay
- Allow drawings to be provided without impacting on the in-house architectural team
- Allow swift implementation of the works with approval of the stores
- Ensure minimal or no disruption to the trading stores
- Determine likely project durations for typical stores sizes (small, medium and large), so stores can be briefed regarding the anticipated in-house disruption
- Ensure all unitary take-offs were correct and deliveries were arranged for the most suitable time slots
- Ensure the required quality of workmanship was achieved and maintained with minimal supervision
- Meet all legislative requirements (health and safety etc.)

2.2 Content, scope and scale

The site works comprised the execution of the following activities at each of the 100 premises:

- Remove all existing wall-mounted displays
- Isolate any unwanted power supplies on the retail walls and fit blanking plates to the back-boxes

- Make good decoration to walls by removing old wall coverings, lining the walls and painting in a new colour finish – in many cases it was necessary to extend the area within the retail space beyond that on which the new displays were being installed

- Alter, where required, the demise lines of the carpet and vinyl tile floor finishes to suit the re-designed layout

- Unpack all new displays – these were manufactured and shipped from China, and as such the packaging comprised plywood crates, cardboard boxes and considerable volumes of polystyrene packaging

- Dispose of all unwanted materials. The old unitary itself would require at least a single skip to dispose of the waste, and the packaging for small- and medium-sized stores would require at least a further skip while large stores could need up to three additional skips. Where practical, the waste was removed from the site using the contractors' own vehicles and taken either to their respective depots for sorting and recycling, or in some cases sent for separation and re-cycling to local waste disposal sites (NB: The unitary generally comprised of timber, metal and Perspex and could easily be separated)

- Install new design wall-mounted displays

- Dismantle and remove all island units (floor displays)

- Assemble and install new island units

- Re-position, where required, any lighting to ensure optimum lighting levels and make good any suspended ceilings

- Undertake a comprehensive clean of the retail area and unitary, so the store staff could re-merchandise the displays the following morning and recommence trading at the earliest opportunity

Prior to these works being implemented, an exercise was required to determine in which 100 stores the work could most effectively be carried out. In addition to the necessary statutory consents, there was an extensive range of internal approval processes, such as individual brand design sign-off on a store-by store basis, financial approvals, landlords' consents and so on.

2.3 Duration

Within the overall 5 month period for completion of the works, typical timescales for individual projects were:

- Start: Two weeks lead-in from receiving all approvals
- Completion:
  - Small store: 1 night
  - Medium store: 1–2 nights (additional labour allowed a single night in some cases)
  - Large store: 2 to 3 nights (the physical size of some stores outweighed the benefit of additional labour)
- Defect inspection: Within 3–4 months from practical completion (PC)
- Payment: 100% made 2 weeks from PC

3 Programme organisation

3.1 Overall Structure and relationships

Overall sponsorship and governance of the programme was provided at an executive level within the retailer's organisation together with joint venture partners who supplied all funding for the works. While programme strategic direction, policies and

Appendices

processes was controlled at the programme level, actual implementation was delegated down to the project level. All projects followed a standard approach for governance, control and reporting mechanisms.

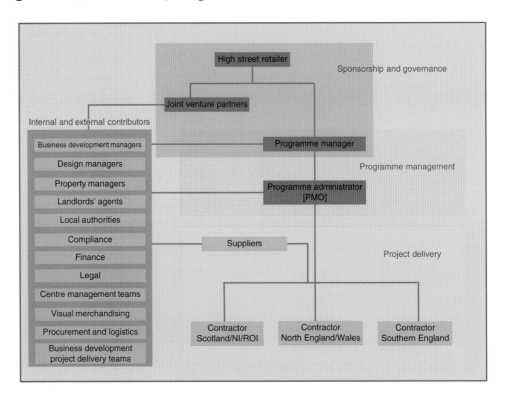

3.2 Roles of key participants and responsibilities

Key Stakeholders

- Main retailer (High Street Brand)

- Joint venture partners (JVPs) – Financial responsibility for all funding of projects on a store-by-store basis

- Retailer/JVP representatives:

    - Programme manager – Overall responsibility for programme delivery

    - Programme administrator (PMO) – Day-to-day co-ordination of programme and projects. Determining high, medium and low priority sites based on various factors (i.e. mall stores have no requirement for advertising consents, any stores known to have asbestos would require further investigation and store likely to be relocated in the next 12–14 months may not be worth progressing, etc)

    - Business development managers – Business to JVP interface

    - Design – Ensuring brand protection and practical operational design

    - Property – Co-ordination with landlords' agents for necessary approvals

    - Compliance – Checking asbestos register and advising of at-risk sites

    - Finance – Ensuring JVP funding is available for subsequent payment of all invoices

    - Legal – drafting of power of attorney document for each project, and overall control of bespoke shopfitting contract with the appointed shopfitters

    - Visual merchandising – Ensuring optimum display of products in each location

    - Procurement – Forecasting of unitary and other goods for manufacturing

- Logistics – Co-ordinating purchase and delivery of all direct supply goods
- Business development project delivery team – Completion of 3 month defects inspections and compliance of 'as built' drawings issued by each contractor. Carried out on a regional basis

External Parties

- Shopfitting contractors (x3) – All onsite works
- Landlords' agents – Approval of proposed works
- Signage manufacturer – Signage design, submission of advertising consents to appropriate local authorities, manufacture and installation of new signage
- Local authority – processing and approval of advertising consents
- Centre management teams – processing and approval of all works within shopping malls
- Unitary manufacturer – Manufacturing and shipping of all unitary from China to UK
- Logistics and distribution – Importing, shipping and site delivery of all unitary and free issue goods to store addresses

4 Programme management process and practices

In this case study the early stages of the programme life cycle relating to inception, initiation and definition had been developed at an executive level in the retailer's organisation. At the end of Stage C the definition of what the programme needed to achieve and the expected benefits were passed to a newly appointed programme manager for implementation.

During a rapid review of potential delivery strategies, it was concluded the only realistic way of achieving the objectives within the required timescale was to adopt a programme management approach to provide centralised direction, control and monitoring of the 100 projects and to embrace a partnering arrangement with existing contractors who were given extended responsibilities and autonomy for project delivery. Contractors were given responsibility for confirming the store layout design, drawing production, site management and supervision, determining project duration and cost, self-snagging and agreeing sign-off with the store.

The process required little induction or training of either the approved contractors, designers or project teams as their historical knowledge and experience of working with the clients was extensive. However their delivery teams were clearly briefed on each aspect of the fit-out to eliminate any doubt as to how the finished product would look. Upon completion of the first two sites for each of the contractors, a quality control inspection was carried out with both the client's lead project manager and contractor's project manager and site supervisors present to ensure all parties understood the acceptable parameters with regard to the finishes and to discuss the general project delivery. To finalise the whole process, a joint meeting with all of the contractors was held upon completion of five stores each. The final agreed process was then passed over to the PMO who had responsibility to ensure this process was rolled out across all the 100 projects. The PMO monitored the progress of each project against this process and ensured any issues preventing contractors from proceeding were rapidly escalated to the programme manager for resolution.

5 Benefits realisation

The principal and expected benefits that were anticipated from adopting the programme management approach were:

- Speed of delivery
- Cost effectiveness – The project format offered notable savings due to the opportunity to negotiate with each shopfitter

The 100 projects were very cost effective, with minimum variations, and the overhead cost to the franchisees and the main group was minimal. The turnaround of the sites successfully met the board's target of 100 stores by the required date, and as a result a further 150 stores were added to the programme.

No major dis-benefits were found. There were unintended benefits that resulted from the programme:

- The new retail appearance generated increased sales more than anticipated. This, along with the speed of delivery, increased store turnover dramatically and, in effect, brought the increase in turnover and profit earlier than a conventional project delivery would have done

6 Lessons learned

Implementation of this programme of works identified a number of factors that were considered important to the successful delivery of further similar programmes or projects:

- Providing open sharing of knowledge across the wider team
- Giving total delegation of tasks to capable individuals and teams
- Placing trust in those individuals to deliver
- Embracing a partnering ethos to implementation
- Establishing a strong client/supplier relationship

Case Study 3 – Example of an Emergent Programme: Highways England

1 Introduction

Highways England (prior to 2015 known as the Highways Agency) is the government body with responsibility for operating, maintaining and modernising the strategic road network in England; all motorways and major A roads totalling 4,300 miles and carrying a third of national traffic. In common with all public bodies, Highways England was required to obtain funding for new capital works by bidding to HM Treasury on a project by project basis with monies being released on an annualised basis. Although a 4 year rolling plan of schemes was put in place, the release of annual funding necessary to achieve this was not secure as often changing priorities of other categories of transport regularly caused peaks and troughs in availability of funds. This lack of certainty of funding created substantial inefficiencies in the delivery of new schemes.

Coinciding with the new coalition government beginning to instigate their austerity programme, Highways England had carried out a strategic business review of their delivery requirements and processes and on the basis of the findings approached HM Treasury with a unique proposal. They made an offer that if HM Treasury would allow them to adopt a programme approach and if they would guarantee availability of funds for this 5 year programme of 14 schemes without the restrictions of annualisation, Highways England would offer a 20% reduction to the cost of delivering the schemes.

The success of the adoption of programme management has resulted in HE promoting a 'programmatic approach' as an organisational culture that they apply to their capital delivery and change management activities.

2 Programme description

2.1 Aims and objectives

Programme goal: To undertake their 2010–2014 Delivery Plan with a 20% reduction in cost.

Highways England had a number of strategic objectives:

- To adopt a programme management approach to the 2010–2014 Delivery Plan to radically increase the effectiveness of delivery
- To achieve savings of at least 20% on the costs of delivery
- To establish a culture of collaboration and openness in delivery

2.2 Content, scope and scale

The programme comprised the delivery of 14 major roadway schemes of a total estimated cost of £1.78 billion.

2.3 Duration/time line

The overall programme of works was to complete within a 5 year period with flexibility in the order and timing of individual projects.

3 Programme organisation

3.1 Overall structure and relationships

On HM Treasury's acceptance of their offer, HE established a programme management organisation headed by a programme director (at divisional director level within HE), with a delivery hub acting as a programme management office as the central driving force of programme delivery.

Appendices

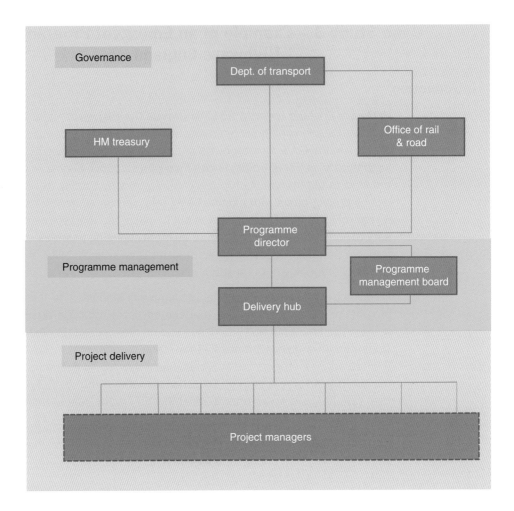

3.2 Roles of key participants and responsibilities

Key roles and participants comprised the following:

- Programme Director: overall responsibility for the successful delivery of the programme of 14 schemes and for securing funding from HM Treasury

- Programme Management Board: a steering group comprised of senior personnel from across the programme that provided overall direction to the programme

- Head of Delivery Hub: responsible for ensuring all policies and processes for the programme were defined and that individual projects proceeded in accordance with these

- Delivery Hub (PMO): team responsible for developing and maintaining policies and processes and for monitoring the development and performance of projects. Also provided an element of technical design for the projects

- Project Managers: prime responsibility for the delivery of one of the 14 individual projects

3.3 Key stakeholders

Key external stakeholders consisted of the following:

- Department of Transport: government ministry responsible for the country's transport infrastructure

- HM Treasury: government funding agency

- Office of Rail and Road: government body that monitors HE's management of the strategic road network

- Local Authorities/Police Forces: critical role in facilitating statutory and regulatory approvals and for communications regarding impact/disruptions caused by works.

- Trade Press/Notable Think Tanks (e.g. RAC Foundation): able to influence perceptions of project success

Key internal stakeholders consisted of the following:

- HE Operations: Body within HE with the responsibility for the ongoing traffic management and maintenance of the road network

- Programme Management Board: steering group for the programme

- Programme Management Delivery Structure: comprising programme director, head of delivery hub and members of the delivery hub

- Projects Delivery Structure: comprising HE project managers, PM consultants, design consultants, contractors and suppliers

- HE Technical/Engineering Functions: holders of design standards

- HE Strategic and Business Planning functions: advisers on HE's strategic direction and policies

4 Programme management processes and procedures

4.1 Key stages

Whereas previously the methodology for carrying out schemes followed a traditional contractual arrangement of having two teams each with its own stage, one for design and one for construction, on adoption of the programme approach a new process involving three stages was introduced.

Product Development

This first stage was focused on understanding the general requirement and features of each scheme, highlighting potential risks and problems, identifying any potential negative/positive interfaces between schemes and consideration of any critical operational characteristics.

Key outputs from this stage were:

- setting of overall delivery objectives

- developing overall sequencing and target deadlines

- developing cost targets

- identifying efficiency savings

This stage highlighted the significant benefits gained from being able to discuss programme-wide requirements and demands with the supply chain rather than being restricted to only being able to consider each scheme independently.

Construction

To take full advantage of the ability to adopt a programme approach it was decided to promote an open, collaborative culture; a framework arrangement with four delivery partners was established for the construction works. Partners were invited to share

Appendices

in both the pain and the gain of the programme. Incentives were introduced whereby a third of any savings to the target cost on a project were retained by the contractor with the rest going into a programme pot. Contractors were therefore directly concerned with the financial performance of every project, leading to the sharing of experiences, techniques and methods.

Testing and commissioning

Historically incorporating a new scheme into the live road network was a separate process carried out by a separate unit within HE, HE Operations. Often during this process problems arose requiring additional modifications and causing delays to the opening of the new roadways. As a consequence of adopting programme management, this unit was included as part of the programme team and was able to review design proposals to eliminate or minimise any issues arising between the construction works and the final operational facility.

4.2 Process and practices

In addition to the three key programme management stages HE also operate on each project within a detailed project control framework (PCF) which consisted of sequential stages 1–7, each with a gateway approval. (For more detail see HE, *The Project Control Framework Handbook*).[1]

Option Phase		Development Phase			Construction Phase	
1 Option identification	2 Option selection	3 Preliminary design	4 Statutory procedures	5 Construction preparation	6 Construction & handover	7 Closedown

The Delivery Hub was responsible for maintaining 14 different functional areas across the programme:

- Risk
- Planning
- Efficiency and Lean Construction
- Health and Safety
- Commercial
- Financial
- Procurement/Category Management
- Visualisation
- Design Development and Departures
- Organisational Development
- Operations
- Communications

---

[1] Highways England, *The Project Control Framework Handbook* can be viewed at 'http://assets.hoghways.gov.u/kour-road-network/managing-our-roads/project-control-framework/.

- Technology

- Governance/Document Control

Implementation and maintenance of these functional areas was carried out by a mix of HE internal staff, consultants and contractors. Aspects of particularly important programme functions are highlighted in the following sections.

Risk: Initial consideration of risk disclosed that in the past risk had been treated in different ways on projects. Adoption of common definitions and process uncovered a high degree of commonality of risk type so that strategies could be developed at the programme level, which could either eliminate or minimise risk or its impact. Considering risk at the programme level allowed for a more cost effective way of dealing with the likelihood of aggregate risk occurring. Overall, across the programme it was estimated that this resulted in between 2% and 6% saving on costs.

Planning: Production of an integrated high-level schedule for the programme provided a wider perspective which helped to introduce a level of flexibility into the sequencing of projects that would not have been possible on a project by project basis. This allowed for more efficient balancing of demand for plant, materials, technology and labour. Developing schedules in conjunction with the supply chain had the additional benefit of their being better informed about future requirements. Discussions on scheduling with open and transparent attitudes being displayed by all parties contributed to a 5% improvement in project durations with a saving on preliminary costs, which on civil engineering projects can represent as much as 45–50% of total cost.

Health and Safety: The inherent nature of highway construction, involving major items of plant, means the industry does have closely monitored health and safety processes. Consideration of health and safety at the programme level meant that good practice could be taken from some projects and shared on others. In addition, a detailed investigation by the Delivery Hub of a category of 'near misses' across projects led to significant reductions in reportable accidents. Overall the whole programme had a very low accident frequency rate.

Visualisation: The procedure, replicated practice from the car industry, of starting each week with a production meeting involving all parties was implemented. Similar to the methodology of the concept of The Last Planner, the meeting considered performance attained in the previous week and the activities to be undertaken in the coming week. This approach provided full transparency of the current issues, identified problems and blockers, and allowed collective solutions to be developed. Open display of performance measurement was provided by visibility boards in the Delivery Hub's offices and gave a highly transparent view of performance against goals.

> The visualisation is about openness and clarity. Problems are identified and allocated to a named individual to progress the solution. The problem is shared openly for others in the team…, not only to visually see progress, but also to offer input and support – a collaborative approach.
> Statement by Highways Agency Delivery Hub from ICG's 'Improving Infrastructure Delivery'

4.3 Collaboration

A critical aspect of HE's adoption of programme management was the desire to introduce a behaviour of open, collaborative working with their supply chain. This was the focus of the organisational development function area and was achieved by entering into framework agreements with preferred suppliers and by involving them at an early

Appendices

stage in the programme, giving them full visibility of the programme's performance and allowing them to contribute towards jointly developed solutions to problems.

A fuller description of HE's approach to collaboration is contained in the Infrastructure Client Group paper 'Improving Infrastructure Delivery: Alliancing Best Practice in Infrastructure Delivery'.[2] This document describes an alliance as:

> an arrangement where a collaborative and integrated team is brought together from across the extended supply chain. The team shares a set of common goals which meet client requirements and work under common incentives.

Successful alliancing is based on four aspects; the *behaviours* displayed by the individuals involved and their organisations need to be sympathetic to objectives of the programme, all parties needed to be *highly integrated*, there needs to be strong and committed *leadership* of the alliance and there are *commercial rewards* in place for delivery performance.

On this programme HE brought together an alliance comprising their internal programme management structure and delivery hub together with their external delivery partners.

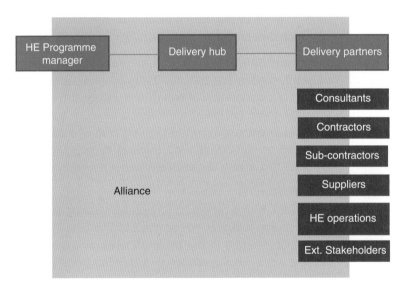

5 Benefits realisation

The principal benefits that resulted from adopting the programme management approach were as follows:

- A significant saving in the costs of schemes providing better value to government expenditure
- Increased speed of delivery
- The development of an open, integrated team culture throughout HE's supply chain
- Better utilisation of resources and reduction in project timescales leading to the savings in preliminary costs

---

[2] Infrastructure Client Group's 'Improving Infrastructure Delivery: Alliancing Best Practice in Infrastructure Delivery' can be viewed at www.gov.uk/government/publications/infrastructure-client-group.

Appendices

- Engagement of a wider programme teaming a smoother transition from construction to operation
- Earlier engagement with the supply chain making them better informed and therefore more prepared for future demands
- A more effective treatment of risk exposure
- Better health and safety, with a dramatically improved accident frequency rate

No major dis-benefits were identified by HE. HM Treasury being impressed by HE's achievement instructed the addition of six further schemes to the 2010–2014 Delivery Plan. Having demonstrated the benefits of using programme management to delivery 5 year plans of work, in agreement with HM Treasury, this has become the preferred method of delivery for capital works at HE

6 Lessons learned

Implementation of this programme of works identified a number of factors which were considered important to the successful delivery of further similar programmes or projects:

- Need for programme leaders to demonstrate their real commitment to openness and trust before this behaviour gets fully taken on board by other team members
- Acknowledging the different objectives of different parties/organisations
- Placing trust and empowerment in individuals/organisations to deliver
- Increased contribution and commitment of participants as a result of their early engagement
- Motivation created by high visibility of performance information
- Benefits from using framework arrangements with the supply chain
- Benefits to supply chain of being informed about future resource demands

Bibliography

This document reflects the best practice guidance available within the sphere of programme management and best endeavours have been used to ensure that appropriate acknowledgements are cited throughout the document. Any inadvertent oversights will be corrected at the earliest possible opportunity.

André, G. (2014) *Change Programme Management: Change and Programme Management: The Helicopter View for Doing Things Better.*

André, G. (2014) *Change Programme Management: Setting the Direction and Defining the Destination.*

André, G. & Barbezieux, A. (2014) *Mapping the Landscape: Engaging with Your Stakeholder Environment and Communicating Effectively.*

APM Programme Management SIG (2007) *Introduction to Programme Management*, APM, Princes Risborough, UK.

Association for Project Management (2007) *Co-Directing Change: A Guide to the Governance of Multi-Owned Projects*, APM, Princes Risborough, UK.

Brown, J.T. (2008) *The Handbook of Program Management: How to Facilitate Project Success with Optimal Program Management.* McGraw-Hill, New York.

Capital Ambition/Excellence in Programme and Project Management (2011) *The public sector programme management approach – release* 4, http://pspmawiki.londoncouncils.gov.uk/index.php/Main_Page (accessed April 2016).

Chartered Institute of Building (CIOB) (2014). *Code of Practice for Project Management for Construction and Development*, 5th edn. Wiley Blackwell, Chichester, 113–119.

Delivering programme outcomes and benefits, http://www.pmis-consulting.com/articles/programme-management/ (accessed April 2016).

Department of Business, Innovation and Skills (BIS) (2010) Guidance for managing programmes: understanding programmes and programme management, https://www.gov.uk/government/uploads/system/uploads/attachment_data/file/31978/10-1256-guidelines-for-programme-management.pdf (accessed January 2016).

Hanford, M. (2004) Program management: different from project management, http://www.ibm.com/developerworks/rational/library/4751.html (accessed July 2015).

Haughey, D. (2001) A perspective on programme management decision support information, version 1.0 project smart: must read news & information about project management, http://www.projectmanage.com/files/program-management-study.pdf (accessed April 2016).

HM Government (2015) Digital build Britain – level 3 building information modelling – strategic plan, http://digital-built-britain.com/DigitalBuiltBritainLevel3BuildingInformationModellingStrategicPlan.pdf (accessed April 2016).

HM Treasure/National Audit Office (NAO) (2010) *Managing complex capital investment programmes using private finance: a current best practice model for departments*, https://www.gov.uk/government/uploads/system/uploads/attachment_data/file/225320/04_ppp_managing_complex_capital_investment_programmes.pdf (accessed April 2016).

ISO (2014) *Working draft – guidance on programme management: ISO TC 258/SC /WG 04*, http://www.iso.org/iso/iso_technical_committee?commid=624837 (accessed April 2016).

ISO 21540 Working Draft – *Guidance on programme management* – ISO TC 258/SC N2014), ISO, Geneva.

Lycett, M., Rassau, A. & Danson, J. (2004) Programme management: a critical review, *International Journal of Project Management*, **22**, 289–299.

Office of Government Commerce (OGC) (2007). *Managing Successful Programmes*, 3rd edn. TSO, London.

Office of Government Commerce (OGC) (2010). *Portfolio, programme and project management maturity model (P3M3)* version 2.1, https://www.gov.uk/government/publications/best-management-practice-portfolio/about-the-office-of-government-commerce (accessed April 2016).

Pellegrinelli, S., Partington, D., Hemmingway, C., Mohdzain, Z. & Shah, M. (2007) The importance of context in programme management – an empirical review of programme practices, *International Journal of Project Management*, **25**, 41–55.

Price Waterhouse Coopers (PwC) (2014) *When will you think differently about programme delivery – 4th Global Portfolio and Programme Management Survey*, https://www.pwc.com/gx/en/consulting-services/portfolio-programme-management/assets/global-ppm-survey.pdf (accessed April 2016).

Potts, K. (2008) *Construction Cost Management: Learning from Case Studies*. Taylor & Francis, http://site.iugaza.edu.ps/kshaath/files/2010/10/Construction_Cost_Management_Learning_from_Case_Studies.pdf (accessed April 2016).

Project Management Institute (2008) *The Standard for Program Management*, 2nd edn. Project Management Institute, Newton Square, Pennsylvania.

Rayner, P. & Reiss, G. (2013) *Portfolio and Programme Management Demystified*, 2nd edn. Routledge, London.

Shehu, Z. & Akintoye, A. (2009) Construction programme management theory and practice: contextual and pragmatic approach, *International Journal of Project Management*, **27**, 703–716.

SAMI Consulting/Experian (2008) *2020 vision – the future of UK construction: a scenario based report by SAMI Consulting – St Andrews Management Institute for Construction Skills*, https://www.citb.co.uk/documents/research/csn%20outputs/2020-vision-future-uk-construction.pdf (accessed April 2016).

Shehu, Z. & Akintoye, A. (2010) Major challenges to the successful implementation and practice of programme management in the construction environment: a critical analysis, *International Journal of Project Management*, **28**, 26–39.

Tam, G.C.K. (2010) The program management process with sustainability considerations, *Journal of Project, Program and Portfolio Management*, **1**, 17–27.

Underwood, J. & Khosrowshahi, F. (2012) ITC expenditure and trends in the UK construction industry in facing the challenges of the global economic crisis, *Journal of Information Technology in Construction*, **17**, 25–42, http://usir.salford.ac.uk/22598/ (accessed April 2016).

Index

appointment, 17, 21, 29–31, 41, 47, 54, 86

BCM *see* business change manager
benefits, 1, 6, 27, 37, 48, 83, 95, 111, 115
benefits management, xv, 42–45, 52, 57, 60, 89, 98–99, 136, 137
benefits profile, xv, 48, 105, 119, 129
benefits realisation manager, xv, 19, 22, 53, 54, 83, 85–87, 95, 97, 98, 136–138
benefits realisation plan, xv, xvii, 3, 17, 22, 48, 54, 103, 106, 107
BIM *see* building information modeling
BRM *see* benefits realisation manager
budget, xvi, 12, 20–22, 34, 35, 42, 49, 51, 54–56, 63, 66–69, 69, 70–72, 78, 91, 134, 139, 146, 148, 149
building information modeling, 4, 32, **34**, 56, 72, 73, 75, 96, 109
business case, **3**, 4, **9**, 17, 21, 22, 37, 39–44, 47, 49, 51, 54, 59, 60, 67, 72, 85, 87, 89–92, 94, 95, 98, 112–114, 121, 131, 132, 134, 136, 137, 140, 151, 152
business change manager, xv, 17, 22, 29, 31, 38–41, 51–55, 57, 60, 86, 87, 89, 92, 95–98, 106–108, 110, 111, 113, 129, 135, 136
business objectives, 27, 33, 41, 73
business partner, xv, 23, 29, 30, 39

change, 4, 27, 38, 49, 83, 95, 110, 115
change control, 53, 60, 62, 65, 68, 72, 88, 91, 132, 140
change management, 6, 10, 12, 31, 34, 57, 59, 65, 66, 72, 91, 98, 122, 124, 151, 158
client, xv, 1, 2, 7, 8, 12, 13, 15, 18–21, 23, 29, 30, 34, 39–41, 48, 52, 53, 55, 57, 66, 75, 79, 85–87, 93, 98, 110, 111, 133, 134, 139, 156, 157, 163
closure, 15, 17, 18, 21, 42, 64, 84, 86, 94, 96, 106, 107, 109–114, 131, 132
commercial management, 78–80, 92, 93, 108
commissioning, 46, 66, 75, 81, 93, 106, 149, 161
communication manager, xvi
competencies, 133–135, 137, 139, 141, 142, 144, 146
construction, 1–2, 4, 7, 8, 11, 13, 18, 34, 59, 66, 68–70, 75, 144, 148, 149, 151, 160–162, 164

cost, xvi, 2, 5, 6, 9–11, 31, 32, 35, 38, 39, 42, 48, 49, 51, 54, 56, 59, 60, 62–72, 74, 75, 78–80, 84, 87, 88, 90, 98, 100, 106, 118–120, 123, 125, 126, 129, 134, 136, 137, 139–140, 142, 144, 146–147, 156–158, 160–163
cost management, 56, 69, 71, 80, 139
customer, xv, 4, 7, 11, 15, 18, 23, 34, **42**, 57, 58, 66, 69, 101, 134, 139, 153

definition, xv, 5–9, 11, 15–18, 28, 39, 43, 47–81, 89–92, 99, 107, 156, 162
deliverable, xv, 2, 5, 11, 18, 36, 39, 41, 51, 54, 57, 59, 67, 73, 77, 78, 86, 87, 92, 94, 95, 116, 123, 135
design management, 151
dis-benefit, xv, 17, 100–107, 118, 129, 157, 164

environmental management, 13, 56, 80–81, 93

feasibility, 38, 39, 43–44, 134
financial management, 68, 68–69, 72, 84, 91, 132, 134, 147
funding, xvi, 2, 15, 18, 20, 22, 23, 35–36, 40, 98, 112, 119, 123, 133, 134, 139, 140, 144, 146, 152, 154, 155, 158, 159
funding arrangements, xvi, 18, 35, 44–45, 55, 68–69, 92

head of the programme management office, 38
health and safety, 12, 13, 56, 61, 80, 87, 88, 93, 112, 153, 161, 162, 164

identification process, 17, 37–38
implementation, xvi, 3, 6, 15, 17, 21, 39, 41, 44, 51, 52, 56, 67, 70, 83–93, 108, 111, 112, 133, 134, 137, 143, 153, 155–157, 162, 164
inception, 15, 17, 27–37, 65, 74, 156
information management, 56, 72–75, 92
initiation, 4, 15, 17, 30, 37–45, 59, 83, 114, 156
integration, 2, 4, 34, 70, 72, 75, 86, 133, 151
issue, xv, xvi, 2, 15, 21, 22, 31, 32, 38, 41, 43, 50, 51, 53, 55, 59–62, 67, 75–77, 79, 83–93, 108, 110–113, 119–120, 123, 127, 132–134, 141, 146, 151, 156, 161, 162

Code of Practice for Programme Management in the Built Environment, First Edition. The Chartered Institute of Building.
© 2016 John Wiley & Sons, Ltd. Published 2016 by John Wiley & Sons, Ltd.

Index

key criteria, 36, 60, 134, 140–142, 145, 146
key participants, 29–31, 39–42, 52–57, 85–88, 97–98, 111–112, *150*, *155*, 159
key roles, 52, 133–138, 159

lessons learned, 40, 62, 84, 107–109, 111, 112, 114, 132, 140, 152, 157, 164

monthly report, 69, 70, **125**

opportunity, xv, xvi, 4, 13, 32, 43, 50, 60–61, 78, 89, 90, 96, 100, 108, 110, 133, 137, 139, 140, 142, 146, 154, 157
organisation structure, 6, 20–21, 29–31, 39–42, 52–58, 84–88, 97–98, 110–112
outcome, xv–vii, 6, 10–13, 17, 18, 22, 23, 27, 29–31, 35, 37, 38, 42, 48, 53, 57, 58, 61, 62, 67, 76, 77, 81, 84, 86, 89, 90, 92, 94, 95, 97–100, 109, 112–114, 116, 117, 123, 126, 133, 135, 136
output, xvi, xvii, 2–4, 10, 11, 17, 18, 21, 35, 47, 48, 51, 57, 59, 60, 67, 77, 81, 83, 84, 86, 89, 92, 94–96, 98, 99, 103–107, 109, 112, 114, 116, 119, 121, 123, 126, 129, 160

PDP *see programme delivery plan*
performance, 4, 5, 13, 15, 17, 49, 60, 64–66, 69, 70, *71*, 77, 79, 80, 83–87, 90, 92–93, 100, 106, 107, 114, 128, 129, 133, 136, 137, 139, 146, 148, 151, 159, 161–164
performance monitoring, 83–84, 90
PMO *see programme management office*
portfolio, xvi, 5–9, 9, 10, 11, 25, *25*, 35, 62, 66, 106, 134, 136, 141, 153
portfolio management, 9, 11, 25, *25*
PrgCM *see programme communication manager*
PrgFM *see programme financial manager*
PrgM *see programme manager*
PrgMB *see programme management board*
PrgS *see programme sponsor*
PrgSB *see programme sponsor board*
procurement, 44, 66, 68, 69, 71, 75, 78–80, 92–93, 134, 140, 144, 155, 161
programme, 1, 6, 27, 37, 48, 83, 95, 111, 115
programme brief, xvi, 14, 17, 36–41, 43, 44, 48, 50, 53, 56, 95, 99, 101, 113, 117–120
programme closure, 15, 17–18, 106, 107, 110, 112–114, 131–138
Programme communication manager, xvi, 22
programme definition, 15, 17, 76, 99
programme delivery plan, xvi, 17, 18, 22, 47, 51–55, 52, 58–60, 72, 84–86, 88–92, 93, 95, 99, 103, 106, 107, 109, 112–114, 117
programme financial manager, xvi, 22, 38, 51, 52, 54–55, 87, 98, 111–112, 139–140
programme financial plan, xvi, 22, 51
programme implementation, 15, 17, 39, 87
programme inception, 15, 17

programme management, 1, 6, 27, 37, 48, 83, 95, 111, 115
programme management board, xvi, 21, 55, 86–88, 112, 159, 160
programme management office, xvi, 17, 21, 22, 38, 52, 55–56, 60, 62, 64, 83, 85, 87, 88, 92, 98, 110, 112, 137, 141, 155, 156, 158, 159
programme management structure, 21–23, 55, 56, 95, 163
programme manager, xvi, 7, 12, 17, 20–22, 31, 33, 38–42, 52–60, 67, 72, 74, 79–81, 85–88, 90, 91, 93, 95, 97, 98, 110–113, 121, 126, 131, 133–136, 139, 141, 155, 156
programme mandate, xvi, 14, 17, 20, 27, 29–31, 36, 37, 39, 41, 50, 101, 117
programme monitor, xvi, 21–23
programme organization structure, 18–24
programme report, 34, 69, 125
programme scope, 14, 59, 86, 116, 149
programme sponsor, xvi, 3, 18, 20–21, 29–31, 38–41, 51–55, 57–59, 67, 84–88, 92, 93, 95, 97, 98, 107, 109–115, 126, 129, 132, 133, 135, 141
programme sponsor board, xvi, 21, 29–31, 38–41, 51, 52, 57, 59, 67, 68, 72, 84–87, 90–92, 95–97, 106, 110, 111, 114, 115, 117, 118, 121, 126, 131, 132
programme timescale plan, xvi, 50
programme vision, 17, 20, 60, 99, 116, 118, 126, 135
project management, 1, 2, 5, 6, 9, 11, 21, 56, 64, 75, 83, 85, 88, 93, 123, 139, 141, 142, 144, 145, 153

quality management, 77, 84, 92, 132, 142

relationships, 13, 23, 29, 39, 44, 48, 49, 52, 57, 58, 62, 64, 67, 74, 76, 84–85, 89, 93, 97, 100, 102, 110, 133, 136, 139, 149, 154–155, 157, 158
reporting, xvi, 22, 25, 33, 34, 52, 55, 56, 62, 65, 66, 68–72, 75, 76, 84, 86, 87, 90, 93, 106, 133, 134, 136, 139, 141, 144, 146, 149, 155
risk, 2, 9, 31, 38, 47, 84, 100, 109, 119
risk management, 12, 34, 48, 56, 60–65, *65*, 67, 75, 88, 108, 123, 124, 142

scope management, 58–60
skills, 6, 10, 20, 21, 30, 34, 114, 133–147
sponsor, xvi, 1, 15, 17, 18, 20–23, 27, 29–31, 38–41, 50–53, 55, 56, 58, 64, 84, 85, 95, 97, 98, 109–111, 117, 120, 126, 133, 135–137, 141, 154
stage outline, 27–29, 37–39, 47–52, 83–84, 95–97, 109–110
stage purpose, 27, 37, 47, 83, 95, 109
stage roles, 29–31, 39–42, 52–57, 85–88, 111–112
stage structure, 29, 39, 84–85, 97, 110
stakeholder management, 21, 49, 75–77, 83, 89, 108
stakeholder map, 24, *24*, 58, 76, 108

stakeholders, xv, xvii, 2, 12, 18, 21, 23–24, *24*, 30, 36, 49, 51, 52, 54, 55, 57, *58*, 58–60, 64, 65, 73, 75–77, 83, 87, 89, 92, 93, 100, 107, 108, 110–114, 122, 129, 132–134, 136, 137, 139, 144, 151–153, 155, 159–160, 163
strategic approach, 33
strategic change, 11, 15, 27, 31–35, 58, 76, 112
strategic objective, xv, xvi, 2, 6–9, 11, 14, 17, 31, 34, *35*, 43, 65, 66, 99, *101*, 104, 117, 148, 153, 158
strategic planning, 11, 31, 32
sustainability, 12, 13, 56, 66, 78, 87, 88
sustainability management, 80–81, 93, 112

transition, xv, xvii, 15, 17, 22, 33, 42, 44, 51–53, 57, 59, 86, 89, 94–108, 110, 111, 113, 114, 124, 135, 164
transition management, 89, 94, 124, 135
transition plan, xvii, 51, 53, 57, 94

vision, xvii, 13–15, 17, 18, 20, 25, 27, 33, 36, 59, 60, 73, 75, 99, 113, 118, 122, 126, 132, 135, 141, 148–152
vision statement, xvi, xvii, 17, 20, 27–31, 48, 115–117, 148, 153